IMISCOE Research Series

Now accepted for Scopus! Content available on the Scopus site in spring 2021.

This series is the official book series of IMISCOE, the largest network of excellence on migration and diversity in the world. It comprises publications which present empirical and theoretical research on different aspects of international migration. The authors are all specialists, and the publications a rich source of information for researchers and others involved in international migration studies. The series is published under the editorial supervision of the IMISCOE Editorial Committee which includes leading scholars from all over Europe. The series, which contains more than eighty titles already, is internationally peer reviewed which ensures that the book published in this series continue to present excellent academic standards and scholarly quality. Most of the books are available open access.

More information about this series at https://link.springer.com/bookseries/13502

Anastasia Christou • Eleonore Kofman

Gender and Migration

IMISCOE Short Reader

 Springer

Anastasia Christou
School of Law
Middlesex University
London, UK

Eleonore Kofman
Middlesex University
London, UK

ISSN 2364-4087   ISSN 2364-4095   (electronic)
IMISCOE Research Series
ISBN 978-3-030-91970-2   ISBN 978-3-030-91971-9   (eBook)
https://doi.org/10.1007/978-3-030-91971-9

This Springer imprint is published by the registered company Springer Nature Switzerland AG
The registered company address is: Gewerbestrasse 11, 6330 Cham, Switzerland

Contents

1 Gender and Migration: An Introduction 1
   1.1   Introduction 1
   1.2   Scope and Aims of the Book 6
   1.3   Organisation of the Book 7
   References 9

2 Gendered Migrations and Conceptual Approaches:
Theorising and Researching Mobilities 13
   2.1   Shifting Analytical Parameters: From Women to Gender
        in Contemporary Migration Studies 13
   2.2   Intersectionalities and Conceptual Approaches
        in Researching Gendered Social Transformations 16
   2.3   Key Conceptual Turns in Migration Studies 20
   2.4   Tackling Theoretical, Methodological and Ethical Issues
        in Research on Gendered Mobilities 23
   References 27

3 Gendered Labour 33
   3.1   Early Studies of Female Labour Migrations 35
   3.2   Care and Social Reproduction 36
   3.3   Understudied Sectors and Gendered Migrant Division of Labour . . . 39
   3.4   Skilled Sectors and Gendered Migrant Division of Labour 43
   3.5   Deskilling and Devaluation 46
   3.6   Conclusion 48
   References 49

4 Transnational Families, Intimate Relations, Generations 57
   4.1   Developing Family Migration 57
   4.2   Transnational Families: Concepts, Generations, Relations 60
   4.3   Transnational Parenting and Childhood 63

4.4   Transnational Intimacies and Sexualities   68
4.5   Conclusion   71
References   71

5   Gendering Asylum   77
    5.1   Emergence of Gendered Perspectives on Forced Migration   78
    5.2   Displacement to Europe   82
    5.3   Vulnerability   85
    5.4   Conclusion   88
    References   89

6   Engendering Integration and Inclusion   95
    6.1   Immigration and Integration: Insights and Debates   96
    6.2   Integration Policies, Gendered Interventions and Outcomes   101
    6.3   Beyond 'Integration'? Activism and Inclusion   107
    6.4   Conclusion   110
    References   111

7   Conclusion   117
    References   121

Chapter 1
Gender and Migration: An Introduction

1.1 Introduction

Why has it been important to incorporate gender relations into our understanding of migration processes and to engender migration research? The need to do so does not only stem from the fact that women globally make up just under half of international migrants. Gender is one of the key forms of differentiation within societies which interacts with other social divisions such as age, class, ethnicity, nationality, race, disability and sexual orientation. The drivers of migration impact on women and men differently. Women and men circulate distinctively, whether it be between rural and urban areas, intra-regionally or globally. Labour markets are often highly segregated and the possibility of women and men crossing borders may also be restricted or opened up through gendered discourses, practices, and regulations governing the right to move and under what conditions. Migration may in turn change gender relations within households and in the community and impact on gendered and sexual identities.

Gendered understandings of international migration emerged slowly in the 1970s and 1980s (Morokvasic, 1975, 1984; Phizacklea, 1983; Simon & Brettell, 1986). The special issue of *International Migration Review* in 1984 was titled 'Women and Migration' and highlighted historical and contemporary dimensions of a neglected issue, namely that of rural-urban and international migration and the incorporation of women into wage labour through labour migrations. Until then, women had been largely ignored in writings on international migration; they had been largely relegated to the home and seen as relatively insignificant economically and politically. As migrants, they were depicted as following men rather than as initiators of migration or moving as independent beings. However the gender blindness of migration studies began to be challenged through the writings of feminist scholars in the 1980s and then some mainstream authors in the 1990s (e.g. Castles & Miller, 1993; Cohen, 1995).

© The Author(s) 2022
A. Christou, E. Kofman, *Gender and Migration*, IMISCOE Research Series,
https://doi.org/10.1007/978-3-030-91971-9_1

Initial studies had focussed on women and migration but by the 1990s there had been a paradigm shift to migration as a *gendered process*, where gender reflected the practices and representations of femininity and masculinity and relationships between women and men (see Chap. 2). Nonetheless, gender continued for many writers to connote women's experiences and lives. In a review of the field, the first handbook on this topic (Willis & Yeoh, 2000) noted that a gender perspective has drawn attention to the significance of the household and its reproductive activities (Truong, 1996), in particular of domestic and sex work. Labour migration, as the focus of much gender and migration, demanded an explanation and highlighted the complexities of migratory movements, their temporalities and circularities.

Poised at the cusp of new developments, the review identified emerging trends such as the diversity among women and men in which gender cut across class, ethnicity, sexuality, age and other social variables, an approach would become more evident with the development of the concept of *intersectionality*, the buzzword of feminist scholarship (Nash, 2008 and Chap. 2). Absence of men and masculinity would not be rectified until males were studied as gendered subjects (Charsley & Wray, 2015; Gallo & Scrinzi, 2016; Pasura & Christou, 2018) (Chap. 2).

Though transnationalism questioned the focus on the bounded nation-state in the 1990s, it remained masculinist until a decade later (Mahler & Pessar, 2001; Pessar & Mahler, 2003). The gender and migration literature also increasingly engaged with theoretical analyses of global inequalities and the counter geographies of globalisation that create new circuits linking the Global South and the Global North, and in which women significantly contribute to household survival in economies destabilised by economic restructuring and withdrawal of public welfare (Sassen, 2000). Concepts such as the global chains of care (Hochschild, 2000) reflected the growing global demand for reproductive labour (domestic, care and sex work). Though family migration had received relatively little attention (Kofman, 2004), with increasing labour migration more families were forced to live apart and stretched across space, as the study of transnational families in Europe (Bryceson & Vuorela, 2002), Asia (Yeoh et al., 2005) and North America (Hondagneu-Sotelo & Avila, 1997) revealed.

Three socio-economic and political changes have also oriented the development of gender and migration. They are firstly the enlargement of the EU and the growth of mobilities and migrations from Eastern to Western and Southern Europe where research has focused on domestic and care labour exemplifying the global chains of care (Lutz, 2011; Marchetti, 2013) as well as family networks (Ryan et al., 2008). The second is the financial crisis, especially severe in Southern European countries which had less impact on migrant women's employment, though it often put additional pressure on them as breadwinners. The loss of employment brought about new mobilities between sending and receiving countries (Herrera, 2012). It also led to emigration from Southern to Northern European countries, but here we know less about its gendered outcomes (Bartolini et al., 2017; Lafleur & Stanek, 2017). Thirdly conflicts in the Middle East and Africa generated large flows of asylum seekers and a renewed interest in gendered aspects of refugee flows and settlement in Europe in academic and policy studies (Freedman et al., 2017; Williams et al., 2020).

Throughout this period in the growth of studies of gender and migration, it has become common to speak of the *feminization of migration*, noted as one of the four key trends in the age of migration (Castles & Miller, 1993). Yet in the last few years, the notion of the recent feminization of migration has been challenged (Donato & Gabaccia, 2015; Schrover, 2013). Donato and Gabaccia (2015) note that globally the percentage of female migrants has only increased by a small amount from 46.7% in 1960 to 49.6% in 2005. They argued that migrations had already begun to feminise in the early twentieth century in settler societies and Europe. In the United States the share of women in immigrant flows increased sharply between the 1830s and 1860s, and again in the first half of the twentieth century, to attain 50% in 1930. In many European countries it was gender balanced before World War II, leading Schrover (2013: 123) to comment that 'if there was ever a period of feminization, it was in this interwar period'.

During the twentieth century, the composition of flows tended to change according to immigration policies, recruitment practices and the nature of the labour market. In the 1920s, a number of countries restricted male migration but allowed female migration. Many German women migrated as domestic workers to the Netherlands and Scandinavian countries (Schrover, 2013: 112). After the war labour shortages emerged by the end of the 1940s. States with colonies, such as France and the UK, had largely free movement within the colonial system, often recruiting women for low level service and welfare work as with Caribbean women in France and Britain (Byron & Condon, 2008). Other sources of labour in the UK came from displaced persons camps and the Baltic (McDowell, 2016). The liberalization of labour flows in Western Europe following the establishment of the European Coal and Steel Community initially favoured men but from the mid-1960s, the growth of the electronics industry and the search for so-called nimble fingers led to the recruitment of female labour in Germany beyond Southern Europe to countries such as Turkey (Erdem & Mattes, 2003). Sectors such as domestic work and concierges were largely filled by Southern European women, as Laura Oso (2005) highlighted for France. It is estimated that until the economic crisis of 1973 that the 'guest worker' recruitment comprised about 70% men.

However, the global average masks substantial differences between regions due in large part to types of migration (Fig. 1.1).

Regions such as Europe and North America, Australia and New Zealand, and Latin America and the Caribbean have a gender balance with a very slight increase since 1990. These regions offer permanent settlement as well as the right to family reunification and the possibility for family members to accompany labour migrants. This tends to push up the gender balance due to the feminised nature of family migration. In contrast, in other regions, male predominance has risen slightly since 1990. In Western Asia, which includes the Gulf States, demand for less skilled and skilled male labour has been strong even though female domestic workers are also in demand (Malhotra et al., 2016). Thus, the number of female migrants may have increased in absolute terms while relatively declining, a distinction which should be

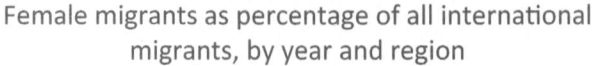

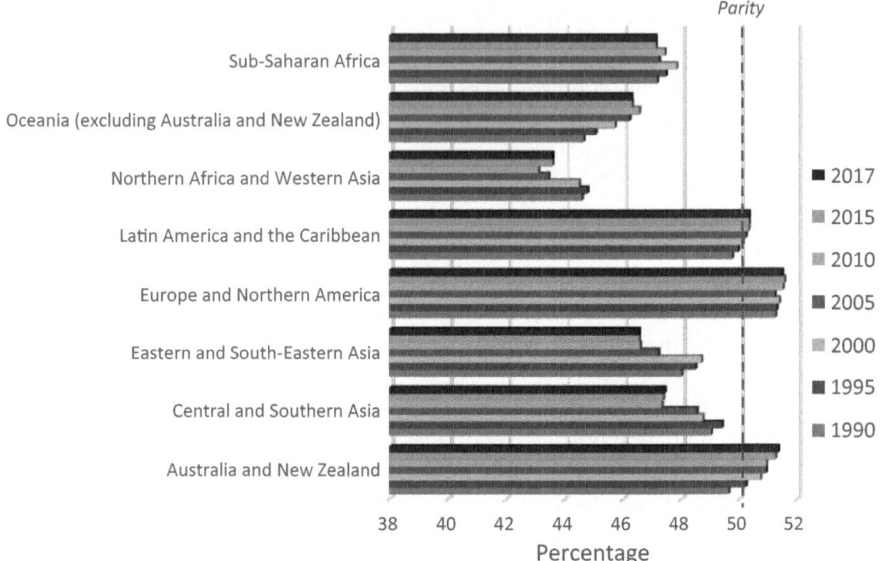

Source: UNDESA (2017)

Fig. 1.1 Proportion of female migrants of all international migrants 1990–2017

taken into account in the discussion on the feminization of migration (Vause & Toma, 2015). The evidence does not substantiate the view that feminization has been linear nor always a new development, but rather that it is dynamic and complex. We need to distinguish between the feminization of migration and the feminization of the 'migratory discourse' in which women are conceptualised as actors of migration (Schrover, 2013; Vause & Toma, 2015).

Lastly, a different critique of the view that migration has become feminised draws upon the increasingly higher levels of education of migrant women to contend that what we have been witnessing in the past few years is the feminization of skilled migration (Dumitru, 2017). Highly educated women in particular are migrating to a much greater extent than men with a similar educational level with the number of tertiary educated migrants increasing by 79% between 2000/01 and 2010/11, 17% greater than for male migrants to OECD countries (OECD, 2016). In the countries of the South, among women aged over 25 years, highly educated women are the most mobile groups, especially from poorer countries, such as sub-Saharan Africa, where almost 20% of the highly educated in 2010 had emigrated compared to 0.4% of the least educated (Dumitru & Marfouk, 2015). Women have thus formed an increasing percentage of skilled migrants (defined by their level of education rather than the occupation they take up after they have migrated). As we shall see in Chap. 3, there has been relatively little research on highly educated women or those undertaking highly skilled jobs post-migration.

It should also be noted that the discussion about feminization focuses on labour migration, yet in 2015 the largest source of permanent migration in OECD countries was family migration, ahead of labour and humanitarian migration, with 38% of migrants entering through this route. Over 50% of this type of flow are women, with 60% in European OECD countries, 57% of sponsored family in Canada and two-thirds in Australia. In some countries with a large number of family permits, such as Canada, the UK and the US, the proportion of accompanying family of other admissions streams, such as the economic, pushes up the proportion of family-related reasons for migration. Most family migrants are spouses, followed by children and parents (OECD, 2017: 125). In some countries where family reunification is not permitted for less skilled migrants, one of the most significant forms of family migration is marriage migration as in Asia (Chung et al., 2016; Constable, 2005). In general, the number of family migrants may fluctuate according to the general level of migration, as in Southern Europe, or due to shifts in immigration policy where governments seek to control this form of migration, often in favour of skilled labour migration, as in Australia in the 1990s (Boucher, 2016).

The focus on labour migration framed the migration of women within a general push-pull model, even if a social dimension was added. Only more recently has a more comprehensive reflection on how female migration to a much greater extent than male might be driven by a desire to escape socially discriminatory institutions and social control. Evidence from the Social Institutions and Gender Index (SIGI), which measures discrimination against women in social institutions, indicates that gender inequalities serve as both a motivating factor and barrier for women's migration (Ferrant & Tuccio, 2015; Ruyssen & Salomone, 2018). On the one hand, women who face discrimination in their country of origin may want to migrate abroad, and may chose destinations where levels of gender discrimination in social institutions are lower than at home. On the other hand, gender discrimination in countries of origin can also prevent women from being able to migrate, when they have onerous family responsibilities, limited access to resources and social networks, little bargaining power or the right to initiate migration themselves. Qualitative research further supports the finding that discrimination is a driver of women's migration. Studies show, for example, that women migrate internally to larger cities, or across country borders, to avoid child, early and forced marriage and other forms of violence against women in the family (Parish, 2017).

And lastly, we should take into account that migration has become complex in its directions and orientations. It is varied in its duration with migrants not necessarily starting out with fixed intentions or what Engbersen et al. (2013) have called 'liquid mobility'. In Europe the opening up of free movement in 2004 in the context of increasingly liberalized and deregulated labour markets has generated large-scale movements from East to West with such migrants often replacing racialized migrants beyond Europe (Favell, 2008). Subsequently the severity of the economic crisis in Southern Europe, loss of employment, especially among youth and the austerity measures drove many highly educated young people to seek employment and opportunities in Northern Europe (Lafleur & Stanek, 2017).

And whilst, as we shall see in Chap. 3, many young Europeans experienced dequalification and deskilling, especially in the initial period of movement, those with recognized cultural capital, and often from a solidly middle class background, are able to enter more smoothly into skilled occupations, for example, as with Spanish migrants in France (Oso, 2020). In this way gender, racialization, class and age have stratified the outcomes of their migratory projects. The ability for European citizens to move with relatively few barriers has also initiated onward migration of new EU citizens of migrant and refugee background (Ahrens et al., 2016; King & Karamoschou, 2019) which breaks down a straightforward relationship between origin and destination country. Gender plays a part in whether the family moves and in the severity of the often precarious experiences of such onward migrants (McIlwaine, 2020).

1.2   Scope and Aims of the Book

There are a number of ways in which one may structure the field of gender and migration which has in the past two decades begun to crystallise into an epistemic community (Kofman, 2020; Levy et al., 2020) as a production of knowledge amongst a network of scholars around certain topics and approaches. Some scholars have focussed on threading an analysis around key perspectives such as intersectionality and transnationalism (Amelina & Lutz, 2019) or integration (Anthias et al., 2013). In this book we trace the emergence of knowledge production of the field in general followed by the key drivers or motives for migration – labour family and asylum/refugees. These are the building blocks of contemporary migration governance, and as categories implemented by states and international organisations, they shape and control the modes of entry open to migrant women and men and also structure the nature of academic outputs. This is not to say that these categories determine migrant lives or that the categories themselves are fixed; they are in fact entangled, articulated and dynamic.

In the past few years, reflections on the construction of categories in migration research and policy have come to the fore (Crawley & Skleparis, 2018; Dahinden et al., 2021; de Haas et al., 2019; Schrover & Moloney, 2013). Throughout the book we acknowledge the fact that the categories we are dealing with have been determined by states and international organisations, and often disseminated by the media. Classifications and categories emerge in a particular social and political context and period; they may evolve in certain instances, whilst in others they remain largely unchanged in a reality that has changed. Within the broader categories, there are numerous issues which demand critical attention. Especially pertinent are definitions of skills and their gendered implications, the notion of the family, usually nuclear and heterosexist for the purpose of entry, and of the refugee more likely to have been displaced through mass movements arising from conflicts rather than the individual (male) figure envisaged by the 1951 Refugee Convention. Our three prevailing categories are those used to classify modes of entry but the bearers of these classifications assume a range of identities in societies post entry. The labour migrant may form a family, the family migrant is increasingly likely to work, and the asylum seeker/refugee seeks to work and reunite with their family or to form a new one. Thus the gendered division of labour draws upon migrants entering through the entire range of categories i.e. those entering through family channels or asylum also participate in the labour market. So too are the categories and the ways they are applied challenged by activists and researchers. Examples are the heterosexist nature of the family in immigration policy which has, in a large number of European countries, recognised same sex and cohabiting couples as constituting families; equally there has been an attempt to inject gender and sexuality-related persecution grounds into the Refugee Convention (see Chap. 5). In terms of participation in a society, the concept of integration too has been subjected to considerable critique (Anthias & Pajnik, 2014) at a time when many states are imposing greater demands on migrants (see Chap. 6).

Thus this book seeks to cover the general development of the field of gender and migration in the past 30 or so years, both in relation to different forms of immigration and post entry insertion into societies. In doing so, we seek to raise debates and explore different and emerging approaches. Intersectionality has become a major concept in gender and migration studies though it struggles to encompass the full range of the interplay of different social divisions (see Chap. 2). Moving on from women to gender, there remains nonetheless a tendency to focus on women, although the need to recognise men and masculinity is being addressed in general and across a range of topics. So too is the relevance of sexuality in migration patterns and outcomes. Most of the literature referred to in the Reader is in English but we acknowledge the large bodies of academic and policy writing in other languages, and in particular French, German and Spanish. Whilst both authors subscribe to looking beyond the global North as the source of theoretical insights as part of the decolonisation of gender and migration and the uneven circulation of knowledge (Fiddian-Qasmiyeh, 2020; Grosfoguel et al., 2015; Kofman, 2020), the restrictions of a short reader have meant the book is largely limited to a European focus unlike in a longer volume (Mora & Piper, 2021). Nonetheless, wherever possible, we have incorporated wider theoretical insights. The limitations imposed by length have also meant we have been unable, except to some extent through the discussion of transnational families, to connect up origin and destination, though here too we do not assume migration is in any way a linear or permanent journey.

1.3   Organisation of the Book

In this Chap. 1 we have briefly traced the development of gender and development from the 1980s and then in the 1990s the adoption of the idea of feminization of migration propounded by mainstream scholars as well. However others have questioned the simplicity of the analysis and suggested gendered patterns as more complex geographically and over time.

In Chap. 2, we turn to the major theoretical perspectives and the shifting analytical parameters from women to gender and the introduction of intersectionalities, said by some scholars to be the key contribution of feminist scholarship. We then examine some of the recent conceptual developments and methodological shifts and their implications for gendered understandings of migration. We end Chap. 2 with a discussion of research and ethics in undertaking migration studies.

In Chap. 3, we turn to one of the major empirical areas of study in gender and migration, that of gendered labour. This title reflects the fact that labour may be derived from a number of sources, ranging from labour migration, family migration, asylum seekers and refugees as well as students, and that it may have regular or irregular status. We argue that the labour market for migrants is heavily gendered both among the lesser and more highly skilled sectors. There has been a tendency to focus on what we have called the *emblematic figure of the female migrant*, that of domestic and care work especially supporting the social reproduction of the household, but as we indicate there are other sectors both in the lesser skilled sectors, such as hospitality, and in the skilled, such as health professionals as well as academia which deserve more attention. There are also a few studies of women in predominantly male sectors such as IT and engineering. We also recognise that migrants are distributed across the labour market but there are few studies to draw upon. Indeed focussing on the sectoral division may mean one loses sight of the trajectories of individual migrants both in relation to deskilling as well as social mobility.

Chapter 4 explores family migration, for a long time understudied and treated as a secondary form of migration in which women followed men. As from the beginning of this century it captured more attention and is the main reason for permanent migration. Furthermore familial reasons generate more moves than labour in intra-European mobility. The interest in the family and familyhood has spawned a growing literature on diverse aspects of transnational families and how migrants have engaged with borders and split lives and separated families. Thus transnational parenting and children have become significant topics as have considerations of cross-border intimacies and sexualities.

In Chap. 5 we discuss another form of mobility and displacement in which women traditionally did not manage to get to European shores to the same degree as men who had greater resources to make often difficult and dangerous journeys. We show how gendered representations played a part in maintaining the binary between the 'there' beyond Europe and 'here' in Europe. Gender-related persecution had difficulty in fitting into the male political figure of the refugee enshrined in the 1951 Refugee Convention and the attempts to incorporate such concerns as well as sexual orientation and gender identity in asylum determination. The second part examines the contemporary 'Migrant/Refugee Crisis' generated by recent and protracted conflicts in South Asia, sub-Saharan Africa, North Africa and Middle East, especially Syria (Freedman, 2016). Although initial flows were male dominated and gave rise to representations of male refugees as cowardly and threatening to European societies, after the summer of 2015 the gender balance shifted towards women. In particular, we engage with the critique of the prioritising of certain asylum categories through the application of the concept of vulnerability by states, the European Union and international organisations.

In Chap. 6 we move to academic critiques and debates about integration and the application of the concept to target certain categories of migrants. It has been recognised that integration fails to take into account class and race (Schinkel, 2018) but we argue that gender considerations have also been absent, yet integration measures and policies have targeted migrant women, too often assumed to have come from backward and patriarchal societies in the Global South and are either reluctant or held back by men from integrating. The second section examines the different gendered discourses applied to integration of women and men in the past 20 years, especially targeting Muslims as disrupters of a modern society. The third section seeks to go beyond integration and how migrants have sought to contest discrimination and lack of rights, especially in the workplace, as well as claims to political subjectivities seeking to recognise them as fully participating members of society.

In the Conclusion we suggest that that it is important to understand the history of gender and migration and the way in which particular issues, such as feminization and intersectionality, have evolved. We end by highlighting the emergence of significant events – the COVID-19 pandemic, Brexit and Black Lives Matter – which have implications for the scholarship of gender and migration and our engagement with broader societal developments.

References

Ahrens, J., Kelly, M., & van Liempt, I. (2016). Free movement? The onward migration of EU citizens born in Somalia, Iran, and Nigeria. *Population, Space and Place, 22*, 84–98.

Anthias, F., & Pajnik, M. (Eds.). (2014). *Contesting integration, engendering migration.* Palgrave.

Anthias, F., Kontos, M., & Morokvacic-Müller, M. (Eds.). (2013). *Paradoxes of integration: Female migrants in Europe.* Springer.

Amelina, A., & Lutz, H. (2019). *Gender and migration. Transnational and intersectional prospects.* Routledge.

Bartolini, L., Gropas, R., & Triandafyllidou, A. (2017). Drivers of highly skilled mobility from Southern Europe: Escaping the crisis and emancipating oneself. *Journal of Ethnic and Migration Studies, 43*(4), 652–673.

Boucher, A. (2016). *Gender, migration and the global race for talent.* Manchester University Press.

Bryceson, D., & Vuorela, U. (Eds.). (2002). *The transnational family: New European frontiers and global networks.* Berg.

Byron, M., & Condon, S. (2008). *Migration in comparative perspective: Caribbean communities in Britain and France.* Routledge.

Castles, S., & Miller, M. (1993). *Age of migration.* Macmillan.

Charsley, K., & Wray, H. (2015). Introduction: The invisible (migrant) man. *Men and Masculinities, 18*(4), 403–423.

Chung, C., Kim, K., & Piper, N. (2016). Preface: Marriage migration in Southeast and East Asia revisited through a migration – Development nexus lens. *Critical Asian Studies, 48*(4), 463–472.

Cohen. (1995). *The Cambridge survey of world migration*. Cambridge University Press.

Constable, N. (2005). *Cross-border marriages: Gender and mobility in transnational Asia*. University of Pennsylvania Press.

Crawley, H., & Skleparis, D. (2018). Refugees, migrants, neither, both: Categorical fetishism and the politics of bounding in Europe's 'migration crisis'. *Journal of Ethnic and Racial Studies, 44*(1), 48–64.

Dahinden, J., Fischer, C., & Menet, J. (2021). Knowledge production, reflexivity, and the use of categories in migration studies: Tackling challenges in the field. *Ethnic and Racial Studies, 44*(4), 535–554.

De Haas, H., Castles, S., & Miller, M. (2019). *Age of migration. International population movements in a modern world* (6th ed.). Red Globe Press.

Donato, K., & Gabaccia, D. (2015). *Gender and international migration*. Russell Sage.

Dumitru, S. (2017). Feminisation de la migration qualifiée: les raisons d'une invisibilité. *Hommes & Migrations, 1317*, 146–153.

Dumitru, S., & Marfouk, A. (2015). Existe-t-il une feminisation de la migration interna- tionale? Féminisation de la migration qualifiée et invisibilité des diplômes. *Hommes et Migrations, 1311*, 31–41.

Engbersen, G. A., Leerkes, I. G.-L., Snel, E., & Burgers, J. (2013). On the differential attachments of migrants from central and Eastern Europe: A typology of labour migration. *Journal of Ethnic and Migration Studies, 39*(6), 959–981.

Erdem, E., & Mattes, M. (2003). Gendered patterns: Female labour migration from Turkey to Germany from the 1960s to the 1990s. In R. Ohlinger, K. Schonwalder, & T. Triadafilipoulos (Eds.), *European encounters. Migrants, migration and European societies since 1945* (pp. 167–185). Ashgate.

Favell, A. (2008). The new face of east-west migration in Europe. *Journal of Ethnic and Migration Studies, 34*(5), 701–716.

Ferrant, G., & Tuccio, M. (2015). *How do female migration and gender discrimination in social institutions mutually influence each other?* OECD (Organisation for Economic Cooperation and Development) Development Centre working paper No. 326. OECD.

Fiddian-Qasmiyeh, E. (2020). Recentering the south in studies of migration. *Migration and Society: Advances in Research, 3*, 1–18.

Freedman, J. (2016). Engendering security at the borders of Europe: Women migrants and the Mediterranean 'crisis'. *Journal of Refugee Studies, 29*, 568–582.

Freedman, J., Kivlicim, Z., & Baklauciglu, K. (Eds.). (2017). *A gendered approach to the Syrian refugee crisis*. Routledge.

Gallo, E., & Scrinzi, F. (2016). *Migration, masculinities and reproductive labour*. Palgrave Macmillan.

Grosfoguel, R., Oso, L., & Christou, A. (2015). Racism, intersectionality and migration studies: Framing some theoretical reflections. *Identities: Global Studies in Culture and Power, 22*(6), 635–652.

Herrera, G. (2012). Starting over again? Crisis, gender, and social reproduction among Ecuadorian migrants in Spain. *Feminist Economics, 18*(2), 125–148.

Hochschild, A. (2000). Global care chains and emotional surplus value. In A. Giddens & W. Hutton (Eds.), *On the edge: Living with global capitalism* (pp. 130–146). Joanthan Cape.

Hondagneu-Sotelo, P., & Avila, E. (1997). 'I'm here, but I'm there'. The meanings of Latina transnational motherhood. *Gender and Society, 11*(5), 548–571.

King, R., & Karamoschou, C. (2019). Fragmented and fluid mobilities: The role of onward migration in the new map of Europe and the Balkans. *Migracijske i etničke teme, 35*(2), 141–169.

Kofman, E. (2004). Family-related migration: A critical review of European studies. *Journal of Ethnic and Migration Studies, 30*(2), 243–262.

Kofman, E. (2020). Unequal internationalisation and the emergence of a new epistemic community: Gender and migration, *comparative*. *Migration Studies, 8*, 36. https://doi.org/10.1186/s40878-020-00194-1

Lafleur, J.-M., & Stanek, M. (Eds.). (2017). *South-north migration of EU citizens in times of crisis*. Springer Open.

Levy, N., Pisarevskaya, A., & Scholten, P. (2020). Between fragmentation and institutionalization: The rise of migration studies as a research field. *Comparative Migration Studies, 8*, 29. https://doi.org/10.1186/s40878-020-00200-6

Lutz, H. (Ed.). (2011). *The new maids: Transnational women and the care economy*. Zed Books.

McDowell. (2016). *Migrant women's voices: Talking about life and work in the UK since 1945*. Bloomsbury.

McIlwaine, C. (2020). Feminised precarity among onward migrants in Europe: Reflections from Latin Americans in London. *Ethnic and Racial Studies, 43*(14), 2607–2625.

Mahler, S. J., & Pessar, P. (2001). Gendered geographies of power: Analyzing gender across transnational spaces. *Identities: Global Studies in Culture and Power, 7*(4), 441–459.

Malhotra, R., Misra, J., & Leal, D. (2016). Gender and reproductive labor migration in Asia, 1960–2000. *International Journal of Sociology, 46*, 114–140.

Marchetti, S. (2013). Dreaming circularity? Eastern European women and job sharing in paid home care. *Journal of Immigrant and Refugee Studies, 11*(4), 347–363.

Mora, C., & Piper, N. (Eds.). (2021). *Handbook of gender and migration*. Palgrave.

Morokvasic, M. (1975). L'immigration féminine en France. *L'Année Sociologique, 26*, 563–575.

Morokvasic, M. (1984). Birds of passage are also women. *International Migration Review, 18*(4), 886–907.

Nash. (2008). Re-thinking intersectionality. *Feminist Review, 89*(1), 1–15.

OECD. (2016). *International migration outlook*. OECD.

OECD. (2017). Chapter 3: A portrait of family migration in OECD countries. In *International migration outlook*. OECD Publishing.

Oso, L. (2005). La réussite paradoxale des bonnes espagnoles de Paris. Stratégies de mobiité sociale et trajectoires biographiques. *Revue Européenne des Migrations Internationales, 21*(1), 107–129.

Oso, L. (2020). Crossed mobilities: The "recent wave" of Spanish migration to France after the economic crisis. *Ethnic and Racial Studies, 43*(4), 2572–2589.

Parish, D. (2017). Gender-based violence against women: Both cause for migration and risk along the journey. *Migration Information Source*. 7 September https://www.migrationpolicy.org/article/gender-based-violence-against-women-both-cause-migration-and-risk-along-journey

Pasura, D., & Christou, A. (2018). Theorizing Black African transnational masculinities. *Men and Masculinities, 21*(4), 521–546.

Pessar, P., & Mahler, S. (2003). Transnational migration: Bringing in gender. *International Migration Review, 37*(3), 812–846.

Phizacklea, A. (Ed.). (1983). *One way ticket: Migration and female labour*. Routledge.

Ryan, L., Sales, R., Tilki, M., & Siara, B. (2008). Social networks, social support and social capital. *Sociology, 42*(4), 672–690.

Ruyssen, I., & Salomone, S. (2018). Female migration: A way out of discrimination? *Journal of Development Economics, 130*, 224–228.

Sassen, S. (2000). Women's burden: Counter-geographies of globalization and the feminization of survival. *Journal of International Affairs, 53*(2), 503–524.

Schinkel, W. (2018). Against 'immigrant integration': For an end to neocolonial knowledge production. *Comparative Migration Studies, 6*, 31. https://doi.org/10.1186/s40878-018-0095-1

Schrover, M. (2013). Feminization and problematization of migration: Europe in the nineteenth and twentieth centuries. In D. Hoerder & A. Kaur (Eds.), *Proletarian and gendered mass migrations* (pp. 103–131). Leiden.

Schrover and Moloney. (2013). *Gender, migration and categorisation. Making distinctions between migrants in Western countries, 1945–2010*. University of Amsterdam Press.

Simon, R. J., & Brettell, C. (Eds.). (1986). *International migration: The female experience*. Roman and Allanheld.

Truong, T. D. (1996). Gender, international migration and social reproduction: Implications for theory, policy research and networking. *Asian and Pacific Migration Journal, 5*(1), 47–52.

UNDESA. (2017). Population division. *International Migration Stock.*

Vause, S., & Toma, S. (2015). Is the feminization of international migration really on the rise? The case of flows from the Democratic Republic of Congo and Senegal. *Population, 70*(1), 29–63.

Willis, K., & Yeoh, B. (Eds.). (2000). *Gender and migration*. The International Library of Studies of on Migration, Edward Elgar.

Williams, L., Coskun, E., & Kaska, S. (Eds.). (2020). In *Women, migration and asylum in Turkey. Developing gender-sensitivity in migration research. Policy and practice*. Palgrave Macmillan.

Yeoh, B., Huang, S., & Lam, T. (2005). Transnationalizing the 'Asian' family: Imaginaries, intimacies and strategic interests. *Global Networks, 5*(4), 307–315.

Chapter 2
Gendered Migrations and Conceptual Approaches: Theorising and Researching Mobilities

2.1 Shifting Analytical Parameters: From Women to Gender in Contemporary Migration Studies

While the repetitive rhetoric of 'discovering' women as active agents in mobility decisions, plans and the execution of such, might have had a major contribution in filling an important lacuna in migration studies literature several decades ago now (Morokvasic, 1984; Kofman, 1999), there are a number of analytical problems with continuing claims that seem to either conflate 'gender' with women or tend to nearly essentialise the 'feminization of migration' in reflecting discursive stereotypes. In the latter case, gendered migration research requires taking on board the historicity and local embeddedness of particular case studies which should clearly frame socio-political and development strategies when conducting studies to understand women migrants and female migration (Cornwall et al., 2008; Dannecker & Sieveking, 2009; Amelina & Lutz, 2019). This perspective becomes clear in the following sections and in the box included in this chapter where we include exemplifications from case studies and our own research findings.

The dominating portrayal has been that of international migrants as young males crossing borders primarily consciously for livelihood reasons, either through documented or undocumented means. Yet, even recently through protracted displacement we continue to realise that there is a clear need as migration researchers to generate more disaggregated data by gender, age and family status as to reflect the complexity of vulnerabilities, mobilities and gendered findings (Kofman, 2019). The notion of the popular, but at the same time quite traditional, representation of international migration viewing female migrants as members of family mobilities in the diasporic sense and by extension as more passive followers than active agents in the global mobility phenomenon, along with victimisation discourses in the migrant sex worker industry (Agustin, 2007), has obscured a wider awareness of the *autonomy* and *subjectivities* of women migrants. The lack of such awareness of the complexities and intricacies of the historicity of countries of origin and

© The Author(s) 2022
A. Christou, E. Kofman, *Gender and Migration*, IMISCOE Research Series,
https://doi.org/10.1007/978-3-030-91971-9_2

destination as well as the embeddedness of social, temporal and biographical parameters (Christou & Michail, 2019) is an added element to why the 'feminization' of the 'age of migration' can appear as an over-inflated generalisation.

Conceptually, for almost three decades (Braidotti, 1992; Altamirano, 1997; Silvey, 2004; Nawyn, 2010) feminist theories have increasingly highlighted their developments and contributions to migration research. Two decades ago, Kofman still exclaimed that: 'Methodologically, we are more equipped than ever to probe the temporal and geographical complexities of individual, household and group itineraries. There no longer is any excuse for the gender blindness of European mainstream research' (1999: 289). And, in the 2000s and currently, migration scholarship now accepts that mobilities are gendered phenomena which by extension require more sophisticated analytical and theoretical tools than 'studies of sex roles and of sex as a dichotomous variable allowed in the past' (Donato et al., 2006: 4).

Such a sophisticated and more complex set of tools mirrors what Mahler and Pessar (2001) term as a 'gendered geographies of power' approach which links the interconnectedness of the temporal, spatial, scaler, biographical and other intersections on the individual and family level in shaping experiences. Gendered power should be seen as a core catalyst to unravel our understanding of those individual and collective experiences. Migration as a gendered and *gendering process* is an important realisation for research into the civil, social and political rights of migrants in any destination society (Szczepaniková, 2006). Gendering perceptions are also characteristic of sustaining hegemonic representations of 'white male breadwinner' categories in the post-war (Western) European context despite migrant women filling in important labour market gaps where local origin women would have otherwise been expected to take on (Kofman et al., 2000: 136).

Following Parreñas (2009), in our interrogation of how gender is constituted in migration studies there is a lack of emphasis on gender as a relation of inequality between women and men, and, thus our research should not solely be pointing to differences in femininities and masculinities. That is, in the study of migratory processes it is important that we understand the *gendering* of such as the study of the emergence of *inequalities* among women and men migrants. More importantly, it is crucial that discussions on gendered migration continue to explore the gendered experiences of men migrants, especially their vulnerabilities, marginalisations, affective channels, personal and family ties in a nuanced approach which embraces critical insights of their inequalities (Pasura & Christou, 2018). The invisibility of the migrant man while reduced in recent literature (Charsley & Wray, 2015), still requires a holistic approach to understanding migration trajectories in tandem with structures impacting migrant identities and life stories, the nuances of migrant agency, the complexity of societal spheres, familial, personal and social relationships. As Wojnicka and Pustułka (2019: 91) assert, 'migration as a process influences the changes in defining, negotiating and performing masculinities, while male migrants create a myriad of migration forms. Stating that migration is a gendered and gendering process has conspicuous consequences for men, women and societies, with the notion of migrants' sex preconditioning our reception of migratory flows'. At the same time, it is crucial to underscore that 'migrant men' should not be conflated into a unified homogeneous group as parameters of intersections and/or the matrices of age, generation, class, race, ethnicity, sexuality, ability, etc. are all generative of multidimensional outcomes for male migrant trajectories and their positionalities within given societies. Added to these are the usual translocal and transnational layers that might also shape their mobilities and identities in host and home countries. In this direction, we see an intersectional and holistic approach to the study of migrant masculinities as being an alternative conceptualisation to the more insular view on 'hegemonic masculinity' (Christou, 2016a).

Transnational approaches to migrant masculinities also deepen understandings of how gendered migrations are constituted and how identities might be reconfigured in the process. In such a process, new forms of hybrid masculinities might emerge and so they are also socially constructed, fluid and transformative. Migrant masculinities can also become translatable to otherwise patriarchal hegemonic versions when men are on the move (Datta et al., 2009; Pasura & Christou, 2018) but also in the messiness of everyday life where identities are constantly reshaped by experiential means (Noble, 2009). Ultimately, components of dysfunctional practices in social and familial relations can bring into conversation the public and private spheres where migrant masculinity can be further shaped by struggles, contradictions and power impacts. The crafting of migrant masculinities is also contingent to the aesthetic labour and reproductive labour demands that shape how they are constructed as gendered subjectivities (Warren, 2016; Fiałkowska, 2019; Gallo & Scrinzi, 2019).

Recent research on immigrant men has also sought to bring them back into the analytical frame 'not as androcentric agents, but as actors with gendered, intersectional social locations imbued with both masculine privilege and social marginality' (Hondagneu-Sotelo, 2017: 112). Migrant men thus can be recipients of the empowerment that masculine privilege brings while at the same time they can be racialised and marginalised when their working class ethnicised status intersects with regimes of social and labour geographies that allow for new hierarchies to emerge. This leads to exciting new theorisations drawing from a Bourdieusian lens on 'ethnic habitus' in how marginalised groups 'construct and perform situated dominant masculinity' (Grosswirth Kachtan, 2019: 1489). Such research demonstrates that in particular ethnocultural settings performances of 'worthy dominant masculinity' leads to the exposure of a separation between social and masculine status thus unveiling masculinities as performative, relational and contextual social practices in specific settings (ibid).

Male migrant masculinity and agency can also be viewed through a feminist lens to analyse the complexities of gendered mobility, familial and gender relationships. Choi (2019) proposes the concept of 'masculine compromise' to explore the material impacts on gender practices within family life, gender identity and gender attitudes. Additional research by Vlase (2018: 195) on men migrant subjectivities reveals how migration not only deeply shapes but extensively transforms masculinities often threatened by life events, lack or complexity of life milestones and other social encounters which prompts migrants to 'discursively resecure their sense of adult-male status by framing their experiences transnationally, in a broader socio-cultural context of both home and host countries'. Discursive and experiential accounts of migrant masculinities point to their malleable aspect but this should not been seen as devoid of agency, conscious planning and individual reconfiguration.

2.2   Intersectionalities and Conceptual Approaches in Researching Gendered Social Transformations

The shift from women to gender in migration studies has also seen the opening toward a more generative and intersectional approach to the research on contemporary gendered mobilities (Näre & Akhtar, 2014). The incorporation of multiple categories relevant to the understanding of intersectional hierarchies can unveil inequalities, relationships and meanings in migration that can inform how gendered identities and roles emerge as shaped by social reproduction, class division, generation and other institutional and structural practices.

Such analytical gaps make the more intersectionally theorised works compelling in providing wider insights into social categories, hierarchies and inequalities (Bastia, 2011; Grosfoguel et al., 2015). More specifically, significant contributions on intersectionality in migration studies have reflected on how we theorise through a 'translocational lens' (Anthias, 2020) in order to address the connections between social divisions and identities and to understand hierarchies/inequalities through modalities of relational, processual and spatio-temporal instances of power.

In explaining the origins of 'intersectionality' and its meaning, it is Kimberlé Crenshaw, a law professor at Columbia University and UCLA who coined the term 'intersectionality' more than three decades ago (1989) to describe the way people's social identities can overlap in instances of inequality and discrimination. The concept has triggered heated debates in academic and public discourse, frequently has been misused in its application to research and theorising, has led to the politicization of the idea, as well as a lasting relevance and re-visiting of its parameters in affirming that inequalities are indeed multidimensional and not uniform.

So, if we aim to condense intersectionality into a succinct definition, we can say that it is an analytic framework that attempts to understand and transcend how interlocking systems of power, oppression and privilege interact, and specifically to address through this lens their combined impact on those who are marginalised and disempowered within a given society.

At its crux, intersectionality theory asserts that multiple forms of oppression, such as those relating to gender, class, ethnicity, race, sexual orientation, disability, age, generation, etc. are not experienced separately, but interact upon and reinforce each other.

In feminist scholarship, intersectionality has been accepted as an approach of major significance to inclusive research while also being criticized as ambiguous (Bilge, 2013; Davis, 2008). The conflation of intersectionality as a theory, a methodological approach and an activist application has further weakened its robustness and these three aspects require disentangling (Bürkner, 2012). By focusing on the axes of social divisions and categorisations in a dynamic way we can locate the experiences of those marginalised and excluded by framing through key multidimensional structural oppressions.

Moreover, it is important to acknowledge that only quite recently we find more critical conversations in the academic literature combining *queer* migration theorisations with sexual citizenship studies and wider human rights based approaches in the gender and sexuality areas of research (Lewis & Naples, 2014; Luibhéid, 2018). Such a conceptual approach makes a major twofold contribution to migration studies: by enhancing the visibility of LGBTQI+ migrants, refugees and asylum seekers (see also Chap. 5) within a combined rights and sexual citizenship rights context, and that we also expand a critical engagement of those intersections of sexuality, gender, culture and migration in challenging essentialisations 'in individual expressions of desire and identity and border politics more generally' (ibid: 911). That is, in *queering* the transnational political economies of mobilities of scale we highlight queer social movements politics and multiscalar citizenship practices (Grundy & Smith, 2005; Lewis, 2013).

Box: Queer Theory

Queer theory emerged in the 1990s from two different stands, the post-structuralist thought of Judith Butler, and lesbian and gay politics. Queer theory problematises assumptions of identity politics that sexuality constitutes a stable identity which informs specific lesbian and gay lifestyles, practices and cultural expressions. Queering renders fluid the categories of sexualities and genders and underscores the artificiality of boundaries. More importantly, it gives visibility to any hierarchical division of particular categories of desires, sexual practices and the intimacy of social subjects, frequently excluded as 'dissident' or 'other'. 'Queer' rather stands against homogenizing and contests normativity, whether such practices descend from hegemonic heterosexual discourses or from mainstream lesbian and gay politics in the framing of identities as sites of 'becoming' and questioning of norms. As a result, in migration studies we might consider a 'queer' approach to research, as a distinct methodological approach that aims to perform an act of 'queering', to de-naturalise taken for granted categories of analysis, even beyond issues of sexuality and gender.

Queer theory in migration studies has been applied as a distinct methodological approach to the study of mobility themes shaped by gender and sexuality in seeking to denaturalise categories of analysis and to make

(continued)

normativity visible. The implications for queer migration scholarship are new opportunities to develop the field in interrogating the politics of queer mobilities, the ambivalent spatialities where these might emerge, the fragmented contexts of diverse societies for queer inclusion and the challenging of border regimes, normative and state violence, racial capitalisms and the carceral geographies of bordering as well as the carceral spectacle of suffering of detention centres and refugee camps.

Queer migration scholarship critically engages heteronormative as well as homonormative arrangements of borders, bodies, desires and movements shaped by the bureaucratic institutions of the neo-liberal nation-state and capitalist discourses (Murray, 2014). While sexualities issues are at the centre of these discourses, interconnections with other intersectional categories and their resulting oppressions, exclusions and marginalisations experienced by mobile social actors such as migrants, refugees, returnees and asylum seekers, are the articulations of such encounters. This is because processes of movement and belonging are amalgamated with dynamics of institutions and social relations where intersectional inequalities emerge. These are analytical opportunities for 'queer intersections' to be further investigated (cf. Manalansan, 2006). For instance, queering migrant representations despite elements of contestation and ambivalence can offer a new repertoire of how queer subjectivities and socialiaties emerge and are reconfigured along fixed notions of 'nationhood' and 'citizenship'. These are important conversations that we need to have and even more critically imperative research that needs to be undertaken to inform policy countering a dangerous mixture of xenophobia, homonationalism, transborder transphobias and toxic binaries of othering difference.

For more information see the open access book volume on 'Queer Migration and Asylum in Europe' edited by Richard Mole (2021) here:

https://www.uclpress.co.uk/products/141641#

On an individual basis, otherwise repressed (non-heteronormative) sexualities and the re-negotiation of ageing masculinities involves a disruption of identities constructed during (return) mobilities that requires deconstructing power relations shaped by ethnonational signifiers (Christou, 2016a). In exploring such gendered personal and social transformations it is important to keep in mind the tensions that might intersect when structural, individual and cultural parameters collide with respect to affective and gendered meanings of migrant identities. Methodologically, hermeneutical phenomenological analyses on empirically grounded studies can address the multiple layers of power in transnational gender relations (ibid). Moreover, since sexual normativity is a core exemplification of power and exclusion, it is revealing to acknowledge the provocation by Carastathis and Tsilimpounidi (2018: 1120–1121) about the 'omnipresence of methodological heteronormativity in the (visual) discourse surrounding the declared "refugee crisis"' (see Chap. 5). They assert: 'All migration politics is reproductive politics. The nation-state project of controlling migration secures the racialised demographics of the nation, understood as a reproducible fact of the social and human body, determining who is differentially included, who is excluded, and who is exalted. Citizenship, illegality, and asylum are often affirmed or rejected as inheritable transitive properties that adhere to a person by virtue of heteronormative (or, more rarely, homonormative) configurations of kinship' (ibid).

As a theoretical and methodological approach to migration research, intersectionality draws on the relationality within social contexts of migrancy, power and inequalities, thus drawing out the complexity of mutually constitutive forms of oppression. Critical social geographers continue to advocate for more inclusion of black feminist intersectional perspectives and for an ethically-driven care of its application. This is so it does not reproduce inadvertently white, masculinist, racist and colonialist perspectives including foci on transnational migration and embodiment (Hopkins, 2019).

Critical theorising in this direction can also surpass the impasse of embracing the political implications of migration as emergent of migrant agency. Such an approach is articulated in a theoretically motivated recent piece by Jonsson (2020) in underscoring the profound political implications of ongoing conceptual turns and methodological shifts combining 'borders' and 'agency' while analysing the interconnections of these theoretical contexts. Namely, this piece by Jonsson highlights that the legacies of colonialism and the significance of protest as political agency are intrinsic to the realisation of democracy and should not be ignored in European societies and histories when it comes to migration.

The importance of intersectionality as an analytical framework in migration studies generates an inclusive understanding of migrants as members of multiple (under/privileged) groups and the barriers they might face. These intersections underscore the inequalities of interconnectedness at a personal and systemic global level. While migration has been theoretically placed within approaches that embrace a broad range of sociocultural, transnational and translocal social fields (Anthias, 2012), one of the frequently missing social categories examined has been consistently 'class' in the corpus of literature on the gendered geographies of mobilities (Kofman, 2004; Cederberg, 2017; Fresnoza-Flot, 2017). Class, is quite often, a neglected social signifier in intersectional analyses, yet, its inclusion fills important gaps in integrating social stratification for a more nuanced intersectional analysis of migrant experiences and positionalities (e.g. Anthias, 2012). A Bourdieusian analysis of class and migration has been prominent in migration studies (Erel, 2010; Oliver & O'Reilly, 2010; also see Chap. 3) and class has historically been a significant social categorisation in the study of gendered understandings of migrant ethnicity (Anthias, 1992).

Unlike research about marginalised and underprivileged groups, Mastoureh Fathi's (2017) research, drawing on intersectionality and class analyses of narratives of Iranian women migrants in the UK, focuses on the social positions of highly skilled professional migrant women who as medics grapple with class subjectivities, social location performativities and social hierarchies. These interconnections point to the complexity of othering processes especially in the interplay of gender, class and racializations which tend to operate in contradictory and multiple ways. This is intriguing since intersectional studies concerning the treatment of more privileged groups has been by and large disregarded. The study reveals two crucial components of intersectional class: the persistence and importance of *power* relations on the one hand and the performative legitimacy of *belonging* in varying social fields.

Issues of power and belonging are also linked to how intersectionality can easily become 'depoliticised' in the direction that Erel et al. (2010) conceptualise multiple oppressions in critical sexuality studies. Their work highlights the often neglected exclusionary effects of the concept and in particular transgender and transsexuality queer discourse. The authors contend that intersectionality has conceptually failed in crafting progressive impulses within the multiculturalism debate and as a result there are methodological implications that point to interconnected webs of power and hierarchies.

In the next section we inform this discussion with further insights on key conceptual turns in migration studies. This section is not meant to be exhaustive in providing an inventory of central conceptual trends and directions in migration studies, but rather it aims to highlight some of the key turning points that shaped its trajectory over the past three decades. This sets the scene for further reflective insights in the final section on theoretical, methodological and ethical issues of significance in research concerning gendered mobilities.

2.3 Key Conceptual Turns in Migration Studies

The study of gender and migration has historically seen the entire migration process perceived as a *gendered phenomenon*, across a variety of spatial and temporal scales and various intersections of individual and family cycles, biographical, historical and national time, the politics and governance of states and capitalist world systems (Donato et al., 2006). Such research insights of gender and migration have not only strengthened the interdisciplinary field of migration studies but have also theorised new research approaches and strategies from qualitative, quantitative and mixed methodological perspectives. In this section we outline some of the key conceptual turns in migration studies, make links to the gendered layers of these, or lack thereof, and, reflect on methodological and ethical issues.

In their historical tour of methodological nationalism Wimmer and Glick Schiller (2002, 2003) contend that the shift towards the study of 'transnational communities' is the last historical conceptualisation in post-war social sciences entering the study of migration. This occurring, as they explain in detail in their 2002 and 2003 joint articles, as an epistemic move to disconnect from methodological nationalism rather than an embrace of new objects of observation for migration studies scholars. Since Chap. 4 discusses key parameters of the transnational approach, here, we only highlight the impetus toward this conceptualisation and the diversions from there to other approaches.

One of the approaches going beyond methodological nationalism according to Biao Xiang's (2016) 'epistemological behaviouralism', might on the one hand, redress the limitation of taking the nation-state as pre-given container of social life for migrants in erasing the conceptual divide that internal and international create, on the other, it creates new problematic thinking in 'behaviouralizng' the migration phenomenon. Remarkably, Xiang addresses both conceptual limitations by interlinking them as 'constituted and constitutive assemblages' (ibid: 669) and while reconfigured by mobilities as an analytical perspective to explore broader social changes.

Another recent call recognising the need to go beyond methodological nationalism (and multiculturalism) to identify epistemological approaches to produce new migration knowledge is that of 'methodological interculturalism' as argued by Zapata-Barrero (2019). The core recognition here is the unit of analysis being 'diversity' itself rather than the migrant, the state, the nation or any other lens that does not take a diverse public culture of justice approach linked to solidarity and cosmopolitanism as its core analytical framing. In this sense, methodological interculturalism aligns well with research foci on race and discrimination and wider migration policy research driven by an anti-racist, equity, justice and solidarity agenda. In other words, it aligns with analytical frameworks that focus on unveiling power and exclusion while putting forward policy agendas embracing diverse, just and equitable societies.

Theoretical constructions of migration pose both epistemological and ethical problems that correlate with methodological nationalism. According to Anderson (2019) the two paradigm shifts of the mobilities turn and methodological transnationalism have been highly generative and that is why Anderson builds on critiques and alternatives in proposing an approach Anderson terms 'methodological denationalism'. This particular conceptualisation builds on a clear distinction: that of 'migrant' and 'citizen' in 'migrantizing the citizen' in understanding exclusions of 'non-citizenship'. Anderson (ibid) draws attention to how global, regional and local institutions and processes impact on the migrant/citizen in underscoring the sensibilities of historicising concepts, theories and practices informing research. Anderson suggests that, 'in this way it has the potential to recover relationalities and interdependence to shed light on the impacts of methodological nationalism beyond the academy and into politics' (p. 7). The latter is the core objective of an approach that migrantizes the citizen as Anderson contends it will enable research to make connections between formal and informal exclusions which are often multiple and go beyond citizenship.

Exclusions and belongings have figured in a variety of conceptual turns but that of the cultural and materialities approach to migration studies has sought to engage with migrant cultures from a material, embodied and identities perspective. The latter do not overlap but find an impetus in how culture and cultural geographies of migration (e.g. cultural and diasporic landscapes, see Christou & King, 2010) can trigger paradigmatic trends in examining the materialities and emotionalities of migration (Christou & Janta, 2019), the affectivities of embodied mobilities (Christou, 2011; Christou & King, 2011), and, how these can shape identities

(King & Christou, 2010). Basu and Coleman (2008) utilised early on the term 'migrant worlds' to firstly acknowledge the materiality of migration, secondly the material consequences of mobilities and thirdly the interconnections of movements of objects and migrants simultaneously. This last third point referring to the binaries and divides between people and things is considered to be one of the biggest 'blind spots' that prevents 'from seeing the full picture and complexity of migration trajectories and pursuit' (Wang, 2016: 2).

These are all phenomenological insights to material and migrant cultures in the study of place-making, identity construction and meaning making in how migrants interact with the world surrounding them, the stuff surrounding them and how sensorial reactions emerge in the mundane experiences of mobile lives. Migrant embodied and sensorial engagements with the materialities of their transnational worlds happen through emotions, living, consuming, interacting with others and objects. Some of these aspects are linked to consumption and lifestyle. As previously, it is important to explore how such experiences and stories of migrant gendered narratives in a psychosocial intersectional perspective articulate implications for gendered migrations (Phoenix & Bauer, 2012).

Further recent conceptualisations include that of the concept of 'lifestyle migration' which Benson and O'Reilly (2016) closely consider in capturing its application as analytical tool and alternative way to thinking about mobilities. Operationalising 'lifestyle migration' points to synergies of movements as practices and how they are understood as meanings in the sociological imagination, including aspiration and the processes that situate them. It is a vibrant field of research which over the years has examined a range of groups and destinations reflecting the definition of lifestyle migration as the movement of 'relatively affluent individuals, moving either part-time or full-time, permanently or temporarily, to places which, for various reasons, signify for the migrants something loosely defined as quality of life' (Benson & O'Reilly, 2009: 621). As a conceptual approach, 'lifestyle migration' can be seen as a driver of migration (or return), having a direct impact in shaping post-movement living and the longer-term implications of settlement and integration as these processes unfold over time. The approach extends beyond the economic and political parameters in shaping the kinds of lives that migrants imagine, experience, plan and dream. This for example has implications in recent calls to problematise the theoretical underpinnings of lifestyle migration and to critically examine the conceptual construction of lifestyle migrants. Recent research has initiated a response to this call by demonstrating how social hierarchies add a depth of dimensions such as class, gender and sexuality interconnected to prisms of 'cosmopolitanism' (Dixon, 2020).

There is an implication of agency here in the imagination of lifestyles but these should not be romanticised in stripping away the impacts of the social geographies, social structures and social constructions which are a part of mobile lives. For instance, lifestyle mobilities can become 'discordant' (Botterill, 2017) when material challenges can lead to a sense of emotional entrapment and immobility when practices, state policy, economic constraints and other insecurities can curtail freedom and destabilise initial enthusiasm. Such outcomes directly point to the *relational* framing of this conceptual approach in situating the experiential aspects of particular case studies and understanding that migrant agency can be confronted by wider structures during the mobility and settlement process. Indeed, we can talk about the intersections of the mobilities paradigm with movement and lifestyle choice as 'lifestyle mobilities' (Cohen et al., 2015) in grasping the complexities of migration, leisure and travel. We understand that the *corporeality* of mobility cannot be easily deconstructed in the traditional binaries of work and leisure, home and away, place and movement, lifestyle and identity. All these patterns of movement and living can become enmeshed or uncoupled as defining avenues to research lifestyle mobilities.

Another key 'turn' being recently re-visited in migration studies is that of the reflexive turn. Its emergence in the social sciences and humanities during the 1980s, especially within ethnographic methodologies and stemming from critical theory and feminist politics and philosophy shaped much thinking in textual, social and cultural theory about the production of knowledge. The complex entanglements of knowledge production along with the 'tyranny of categories' (Stewart, 2015) and multiple contestations emerging in decolonial thought recently has posed renewed challenges to migration scholars to question epistemic communities (see Kofman, 2020 for a current intervention on gender and migration). Concomitantly, others have recently argued that reflexivity among research participants should be incorporated in a renewed definition of reflexivity as mutually constitutive by both researchers and research participants in the co-construction of knowledge (Dahinden et al., 2020). Amelina (2020) also inspired by the reflexive turn, underscores the need to develop a 'doing migration approach' that will embrace mobility, immobility and intersectionalities in discursive knowledge and other configurations in practices which are generative of social orders of migration and how we situate migrants conceptually. Amelina sees potential in this analytical approach as being beyond reflexive and 'doing' migration which leads to a conceptualisation of a social production of the migrant in paying attention to performativities of routines, knowledge and power involved in these processes. This clearly connects with the earlier 'gendered geographies of power' approach advanced by Mahler and Pessar (2001) two decades ago in understanding how a gendered optic in transnational mobilities plays a crucial role in the creation, transformation and fortification of transnational social spaces.

In the next section we round up this discussion with further insights on issues of power and hierarchies in research on gendered mobilities highlighting theoretical, methodological and ethical considerations.

2.4   Tackling Theoretical, Methodological and Ethical Issues in Research on Gendered Mobilities

Migration as a phenomenon and its topical themes have been researched from a variety of humanities and social science disciplines (sociology, geography, economics, demography, anthropology, media and film studies, cultural and literary studies, etc.) through multi-method approaches in investigating regional, urban, suburban and rural locations, primarily dominated by a focus in the Global North with some recent explanatory shifts to the Global South and through a number of schools of thought.

Migration studies has benefited from interdisciplinary explorations with qualitative, quantitative, mixed methods, ethnographic, digital and other tools integrated in small and large-scale single case study or comparative studies. It appears that the sky is the limit in studying mobilities, with generations now of trained migration scholars at the postgraduate and doctoral level at a burgeoning number of academic programmes and through a flourishing of funded studies globally.

It is undeniable that migration studies has developed in multi-faceted directions over a number of decades, incorporating different thematic and topical foci, conceptualisations and methodological directions. The shift to acknowledging the Global South as an important angle in the knowledge production within mobility studies (Fiddian-Qasmiyeh, 2020) has been important and, so has, as mentioned previously, the gendered, feminist and queer lens. What a recent mapping of migration studies, through an empirical analysis of the research field, shows is that a recovery of topical connectedness points to a 'theoretical and conceptual coming of age of migration studies' (Pisarevskaya et al., 2019: 1).

Indeed, pushing the boundaries in liberating our analyses from concepts and instead leaping to the realisation that conceptual heuristic devices might probably be 'impositions' as perhaps the body of scholarship of migration studies is itself, can be frightening but also freeing analytically. For one, freedom from clearly defined conceptual devices can possibly become a productive immersion into new theoretical terrains in the tradition of grounded theory often misused and misunderstood in research. Such an inductive approach enables researchers to seek out and conceptualise social phenomena through processes of interpretative and comparative deconstructing of data.

Yet, it is important to acknowledge existing conceptual and theoretical boundaries and not conflate them unnecessarily. For instance, we would disagree with Stephanie Nawyn's (2010: 749–750) assertion stating that: 'Because most people studying gender and migration use a feminist analytic lens, feminist methods, I often refer to gender and migration scholars as feminist migration scholars'. We would avoid such a conflation of gender with feminist approaches as one does not pre-suppose the use of the other. So, for instance, while in studies of diasporas, tourism mobilities, youth migrants, and masculinities we encounter a feminist lens in researching gendered issues (Christou, 2016a, b; Pasura & Christou, 2018; Janta & Christou, 2019), in others we clearly do not see *feminist* theorisations in analysing gender (Kanaiaupuni, 2000; Liversage, 2009; King et al., 2013). Hence, it might appear self-explanatory, but obviously it is not, so remains important to avoid fixed assumptions that, either gendered migration research refers to women only or that gendered migration research is at the same time feminist in its approach, methods and conceptualisation.

As migration is a phenomenon deeply connected to relationships, emotions, experiences, etc. of the lived experience, biographical, narrative and life history methods are a means of capturing the complex, multi-layered and nuanced understandings of migrant lives. The migrant life story can become a way to learn from in how wider issues of social policy can be addressed. This is equally important in social policy research with refugees and asylum seekers where biographical methods can expand the theorising and understanding of asylum (O'Neill & Harindranath, 2006). Narrative inquiry is a means by which we systematically gather, analyse, and represent migrant stories as told by migrants themselves, which challenges more traditional and modernist views of truth, reality, knowledge and personhood. Subjective meanings and a sense of self and identity are negotiated as the stories unfold, bearing in mind that stories are re/constructions of the migrant experience, remembered and told at a particular point in migrants' lives, to a particular researcher and for a particular purpose.

This all has a bearing on how the stories are told, which stories are told and how they are presented and interpreted. They do not represent 'life as lived' but our representations of those lives as told to us. Hence, stories can be viewed as socially situated knowledge constructions in their own right that value messiness, difference, depth and texture of experiential life. In narrative analysis the emphasis is on co-construction of meaning between the researcher and participants. As researchers while being involved in listening we are in a sense 'translating' the conversations. Researchers take in what is being said and compare it with their personal understandings, without filling in any gaps in understanding with 'grand narratives', but rather inquiring about how pieces of the stories make sense together. The process of 'data gathering' and 'analysis' therefore becomes a single harmonious and organic process.

Recent theoretical dialogues on migration in China (Zhu & Qian, 2020) highlight migrant identity and subjectivity as an evolving process of reconstruction shaped by scales, spaces, encounters, interactions and practices unfolding in everyday experiences where 'the purpose of highlighting banality and everydayness is beyond merely providing thick descriptions of migratory experiences but to demonstrate that proper conceptualisations of migration must exceed pure economic reason or a singular rationality. Migration is always complicated by needs, aspirations, interests and pursuits that are situated, unpredictable and contingent on the immediate milieus of movement and encounter. Whether migrants have resources and abilities to adapt to such contingencies affect their wellbeing in profound ways' (ibid 13–14). These processual approaches give insight to subjective nuances which are revealing of migrant life stories and there has been a rich and prolific literature in this domain (Eastmond, 2007; Gómez-Estern & de la Mata Benítez, 2013; Fathi, 2017; Baran, 2018).

Stories of mobility can also be stories of trauma, displacement and exclusion culminating in recent years in a proliferation of new research in this area linked to wider geopolitical developments in the last decade. In this context, migrant children and refugee youth have come to be constructed as 'crisis figures' in Europe (Lems et al., 2020) but such a sense of exceptionality can trigger conceptually flawed categorisations that essentialise accompanied minors (see Chap. 5). Any research requires a process of ethical reflexivity but research with vulnerable populations raises several ethical questions in upholding research integrity (Akesson et al., 2018). Ethnographically driven research with such populations can focus on how research participants make sense of some of the ascriptions and social categorisations applied to them.

It is important to note that research with unaccompanied youth also illustrates that young mobiles are frequently struggling with how the ambiguity of their refugee figurations are portrayed as 'deserving' or 'undeserving' (Wernesjö, 2020). Mobile youth narratives of un/deservingness articulate how they are constituted as social subjects in defending themselves as hard-working, responsible, diligent and not threatening or victimised (ibid). These processes of narration can be seen as 'manifestations of conditional belonging' (ibid: 389) contingent on negotiations of deservingness.

While research into vulnerable migrant groups can be revealing and of particular ethical significance to inform policy and teaching, as researchers we often find that procedural ethics might tend to be overly burdensome, lengthy, irrelevant and, at times paternalistic, when it comes to vulnerable groups. Research taking place in volatile settings following war, disasters, environmental displacement, etc. can trigger much onerous ethical processes which might delay and even hinder research with populations in those areas. In a sense, this is stripping them from a voice to sharing their stories with the rest of the world and hinder more participatory action research or critical ethnographies to be carried out.

An ethical and reflexive approach should inform the entire research process. This begins with obtaining formal institutional approval, subsequently establishing networks and relationships with potential participants and/or gatekeepers to negotiate access, the collection of data, interpretation, analysis, writing and representation. These processes of reflexive ethical practice are akin to micro-ethical perspectives that require researchers being attentive to managing the dis/comfort zones, affective encounters and boundaries to harm (Akesson et al., 2018). Conducting ethical research with migrant populations at large, and more vulnerable in particular, requires adherence to all these procedural and practical safeguarding steps.

Finally, although it has been pointed out repeatedly, the advice of more conscious efforts to decrease the divide between qualitative and quantitative methodologies is beneficial to the field of migration studies in its entirety, but especially in the analysis and theorising of gender in migration studies (Donato et al., 2006). Yet, this remains one of the challenges of furthering analytical approaches of gender and migration, often a tension that could be harnessed in capturing rich insights (see Chap. 5 for the need to disaggregate data by gender and other social divisions) of the situational and relational categories we explored in this chapter. Quantitative data can also be useful in developing comparative perspectives as with the prevalence and types of transnational families and parenting (see Chap. 4).

References

Agustin, L. (2007). *Sex at the margins: migration, labour markets and the rescue industry.* Zed Books.

Akesson, B., Hoffman "Tony", D. A., El Joueidi, S., & Badawi, D. (2018). "So the world will know our story": Ethical reflections on research with families displaced by war [59 paragraphs]. *Forum Qualitative Sozialforschung/Forum: Qualitative Social Research, 19*(3), Art. 5. https://doi.org/10.17169/fqs-19.3.3087

Altamirano, A. T. (1997). Feminist theories and migration research: Making sense in the data feast? *Refuge: Canada's Journal on Refugees, 16*(4), 4–8.

Amelina, A. (2020). After the reflexive turn in migration studies: Towards the doing migration approach. *Population, Space and Place,* e2368. https://doi.org/10.1002/psp.2368

Amelina, A., & Lutz, H. (2019). *Gender and migration: Transnational and intersectional prospects.* Routledge.

Anderson, B. (2019). New directions in migration studies: Towards methodological de-nationalism. *Comparative Migration Studies, 7,* 1–13.

Anthias, F. (1992). *Ethnicity, class, gender and migration: Greek Cypriots in Britain.* Gower.

Anthias, F. (2012). Transnational mobilities, migration research and intersectionality: Towards a translocational frame. *Nordic Journal of Migration Research, 2*(2), 102–110.

Anthias, F. (2020). *Translocational belongings: Intersectional dilemmas and social inequalities.* Routledge. https://doi.org/10.4324/9780203730256

Baran, D. (2018). Narratives of migration on Facebook: Belonging and identity among former fellow refugees. *Language in Society, 47*(2), 245–268.

Bastia, T. (2011). Migration as protest? Negotiating gender, class, and ethnicity in urban Bolivia. *Environment and Planning A: Economy and Space, 43*(7), 1514–1529.

Basu, P., & Coleman, S. (2008). Introduction: Migrant worlds, material cultures. *Mobilities, 3*(3), 313–330.

Benson, M., & O'Reilly, K. (2009). Migration and the search for a better way of life: A critical exploration of lifestyle migration. *The Sociological Review, 57*(4), 608–625.

Benson, M., & O'Reilly, K. (2016). From lifestyle migration to lifestyle in migration: Categories, concepts and ways of thinking. *Migration Studies, 4*(1), 20–37.

Bilge, S. (2013). Intersectionality undone. Saving intersectionality from feminist intersectionality studies. *Du Bois Review: Social Science Research on Race, 10*(2), 405–424.

Botterill, K. (2017). Discordant lifestyle mobilities in East Asia: Privilege and precarity of British retirement in Thailand. *Population, Space and Place, 23*(5), e2011. https://doi.org/10.1002/psp.2011

Braidotti, R. (1992). The exile, the nomad, and the migrant: Reflections on international feminism. *Women's Studies International Forum, 15*(1), 7–10.

Bürkner, H.-J. (2012). Intersectionality: How gender studies might inspire the analysis of social inequality among migrants. *Population, Space and Place, 18*(2), 181–195.

Carastathis, A., & Tsilimpounidi, M. (2018). Methodological heteronormativity and the "refugee crisis". *Feminist Media Studies, 18,* 1120–1123.

Cederberg, M. (2017). Social class and international migration: Female migrants' narratives of social mobility and social status. *Migration Studies, 5*(2), 149–167.

Charsley, K., & Wray, H. (2015). Introduction: The invisible (migrant) man. *Men and Masculinities, 18*(4), 403–423.

Choi, S. Y. P. (2019). Migration, masculinity, and family. *Journal of Ethnic and Migration Studies, 45*(1), 78–94.

Christou, A. (2011). Narrating lives in (e)motion: Embodiment and belongingness in diasporic spaces of home and return. *Emotion, Space and Society, 4,* 249–257.

Christou, A. (2016a). Ageing masculinities and the nation: Disrupting boundaries of sexualities, mobilities and identities. *Gender, Place and Culture: A Journal of Feminist Geography.* https://doi.org/10.1080/0966369X.2015.1058760

Christou, A. (2016b). 'The wretched of Europe': Greece and the cultural politics of inequality. *Humanity and Society.* https://doi.org/10.1177/0160597616664169

Christou, A., & King, R. (2010). Imagining 'home': Diasporic landscapes of the Greek second generation. *Geoforum, 41,* 638–646.

Christou, A., & King, R. (2011). Gendering diasporic mobilities and emotionalities in Greek-German narratives of home, belonging and return. *Journal of Mediterranean Studies, 20*(2), 283–315.

Christou, A., & Janta, H. (2019). The significance of things: Objects, emotions and cultural production in migrant women's return visits home. *The Sociological Review, 67*(3), 654–671.

Christou, A., & Michail, D. (2019). Post-socialist narratives of being, belonging and becoming: Eastern European women migrants and transformative politics in an era of European crises. *New Formations: A Journal of Culture, Theory & Politics, 17,* 70–86.

Cohen, S. A., Duncan, T., & Thulemark, M. (2015). Lifestyle mobilities: The crossroads of travel, leisure and migration. *Mobilities, 10*(1), 155–172.

Cornwall, A., Harrison, E., & Whitehead, A. (Eds.). (2008). Gender myths & feminist fables. In *The struggle for interpretive power in gender and development.* Blackwell Publishing.

Crenshaw, K. (1989). Demarginalizing the intersection of race and sex: A black feminist critique of antidiscrimination doctrine, feminist theory, and antiracist politics. *University of Chicago Legal Forum, 14,* 538–554.

Dahinden, J., Fischer, C., & Menet, J. (2020). Knowledge production, reflexivity, and the use of categories in migration studies: Tackling challenges in the field. *Ethnic and Racial Studies.* https://doi.org/10.1080/01419870.2020.1752926

Dannecker, P., & Sieveking, N. (2009). *Gender, migration and development: An analysis of the current discussion on female migrants as development agents.* COMCAD Arbeitspapiere – Working papers, No. 69. COMCAD – Center on Migration, Citizenship and Development.

Datta, K., McIlwaine, C., Herbert, J., Evans, Y., May, J., & Wills, J. (2009). Men on the move: Narratives of migration and work among low-paid migrant men in London. *Social and Cultural Geography, 10*(8), 853–873.

Davis, K. (2008). Intersectionality as buzzword: A sociology of science perspective on what makes a feminist theory successful. *Feminist Theory, 9*(1), 67–85.

Dixon, L. (2020). Gender, sexuality and lifestyle migration: Exploring the impact of cosmopolitan place-marketing discourses on the post-migratory experiences of British women in Spain. *Current Sociology, 68*(3), 281–298.

Donato, K. M., Gabaccia, D., Holdaway, J., Manalansan, M., & Pessar, P. R. (2006). A glass half full? Gender in migration studies. *International Migration Review, 40*(1), 3–26.

Eastmond, M. (2007). Stories as lived experience: Narratives in forced migration research. *Journal of Refugee Studies, 20*(2), 248–264.

Erel, U. (2010). Migrating cultural capital: Bourdieu in migration studies. *Sociology, 44*(4), 642–660.

Erel, U., Haritaworn, J., Rodríguez, E. G., & Klesse, C. (2010). On the depoliticisation of intersectionality talk: Conceptualising multiple oppressions in critical sexuality studies. In Y. Taylor, S. Hines, & M. E. Casey (Eds.), *Theorizing intersectionality and sexuality* (pp. 56–77). Palgrave Macmillan.

Fathi, M. (2017). *Intersectionality, class and migration: Narratives of Iranian women migrants in the U.K.* Palgrave Macmillan.

Fiałkowska, K. (2019). Remote fatherhood and visiting husbands: Seasonal migration and men's position within families. *Comparative Migration Studies, 7,* 2. https://doi.org/10.1186/s40878-018-0106-2

Fiddian-Qasmiyeh, E. (2020). Introduction Recentering the south in studies of migration. *Migration and Society: Advances in Research, 3,* 1–18.

Fresnoza-Flot, A. (2017). Gender- and social class-based transnationalism of migrant Filipinas in binational unions. *Journal of Ethnic and Migration Studies, 43*(6), 885–901.

Gallo, E., & Scrinzi, F. (2019). Migrant masculinities in-between private and public spaces of reproductive labour: Asian porters in Rome. *Gender, Place and Culture, 26*(11), 1632–1653.

Gómez-Estern, B. M., & de la Mata Benítez, M. L. (2013). Narratives of migration: Emotions and the interweaving of personal and cultural identity through narrative. *Culture & Psychology, 19*(3), 348–368.

Grosfoguel, R., Oso, L., & Christou, A. (2015). 'Racism', intersectionality and migration studies: Framing some theoretical reflections. *Identities: Global Studies in Culture and Power, 22*(6), 635–652.

Grosswirth Kachtan, G. D. (2019). Challenging hegemonic masculinity by performance of ethnic habitus. *Gender, Work and Organization, 26*, 1489–1505.

Grundy, J., & Smith, M. (2005). The politics of multiscalar citizenship: The case of lesbian and gay organizing in Canada. *Citizenship Studies, 9*(4), 389–404.

Hondagneu-Sotelo, P. (2017). Place, nature and masculinity in immigrant integration: Latino immigrant men in inner-city parks and community gardens. *NORMA, 12*(2), 112–126.

Hopkins, P. (2019). Social geography I: Intersectionality. *Progress in Human Geography, 43*(5), 937–947.

Janta, H., & Christou, A. (2019). Hosting as social practice: Gendered insights into contemporary tourism Mobilities. *Annals of Tourism Research, 74*, 167–176.

Jonsson, S. (2020). A society which is not: Political emergence and migrant agency. *Current Sociology*. https://doi.org/10.1177/0011392119886863

Kanaiaupuni, S. (2000). Reframing the migration question: An analysis of men, women, and gender in Mexico. *Social Forces, 78*(4), 1311–1347.

King, R., & Christou, A. (2010). Cultural geographies of counter-diasporic migration: Perspectives from the study of second-generation 'returnees' to Greece. *Population, Space and Place, 15*(2), 103–119.

King, R., Mata-Codesal, D., & Vullnetari, J. (2013). Migration, development, gender and the 'black box' of remittances: Comparative findings from Albania and Ecuador. *Comparative Migration Studies, 1*, 69.

Kofman, E. (1999). Female 'birds of passage' a decade later: Gender and immigration in the European Union. *International Migration Review, 33*(2), 269–299.

Kofman, E. (2004). Gendered global migrations. *International Feminist Journal of Politics, 6*(4), 643–665.

Kofman, E. (2019). Gendered mobilities and vulnerabilities: Refugee journeys to and in Europe. *Journal of Ethnic and Migration Studies, 45*(12), 2185–2199.

Kofman, E. (2020). Unequal internationalisation and the emergence of a new epistemic community: Gender and migration. *Comparative Migration Studies*. https://doi.org/10.1186/s40878-020-00194-1

Kofman, E., Phizacklea, A., Raghuram, P., & Sales, R. (2000). *Gender and international migration in Europe: Employment, welfare, and politics*. Routledge.

Lems, A., Oester, K., & Strasser, S. (2020). Children of the crisis: Ethnographic perspectives on unaccompanied refugee youth in and en route to Europe. *Journal of Ethnic and Migration Studies, 46*(2), 315–335.

Lewis, R. (2013). Deportable subjects: Lesbians and political asylum. *Feminist Formations, 25*(2), 174–194.

Lewis, R. A., & Naples, N. A. (2014). Introduction: Queer migration, asylum, and displacement. *Sexualities, 17*(8), 911–918.

Liversage, A. (2009). Vital conjunctures, shifting horizons: High-skilled female immigrants looking for work. *Work, Employment and Society, 23*(1), 120–141.

Luibhéid, E. (2018). Heteronormativity: A bridge between queer migration and critical trafficking studies. *Women's Studies in Communication, 41*(4), 305–309.

Mahler, S. J., & Pessar, P. (2001). Gendered geographies of power: Analyzing gender across transnational spaces. *Identities: Global Studies in Culture and Power, 7*(4), 441–459.

Manalansan, M. (2006). Queer intersections: Sexuality and gender in migration studies. *International Migration Review, 40*(1), 224–249.

Mole, R. C. M. (Ed.). (2021). *Queer migration and asylum in Europe*. UCL Press.

Morokvasic, M. (1984). Birds of passage are also women. *The International Migration Review, 18*(4), 886–907.

Murray, D. A. (2014). The (not so) straight story: Queering migration narratives of sexual orientation and gendered identity refugee claimants. *Sexualities, 17*(4), 451–471.

Näre, L., & Akhtar, P. (2014). Gendered mobilities and social change – An introduction to the special issue on gender, mobility and social change. *Women's Studies International Forum, 47*, 185–190.

Nawyn, S. J. (2010). Gender and migration: Integrating feminist theory into migration studies. *Sociology Compass, 4*(9), 749–765.

Noble, G. (2009). 'Countless acts of recognition': Young men, ethnicity and the messiness of identities in everyday life. *Social & Cultural Geography, 10*(8), 875–891.

O'Neill, M., & Harindranath, R. (2006). Theorising narratives of exile and belonging: The importance of biography and ethno-mimesis in 'understanding' asylum. *Qualitative Sociology Review, II*(I) Retrieved August, 2020 http://www.qualitativesociologyreview.org/ENG/archive_eng.php

Oliver, C., & O'Reilly, K. (2010). A Bourdieusian analysis of class and migration: Habitus and the individualising process. *Sociology, 44*(1), 49–66.

Parreñas, R. (2009). *Inserting feminism in transnational migration studies*. http://lastradainternational.org/doc-center/2197/inserting-feminism-in-transnational-migration-studies

Pasura, D., & Christou, A. (2018). Theorizing Black African transnational masculinities. *Men and Masculinities, 21*(4), 521–546.

Phoenix, A., & Bauer, E. (2012). Challenging gender practices: Intersectional narratives of sibling relations and parent–child engagements in transnational serial migration. *European Journal of Women's Studies, 19*(4), 490–504.

Pisarevskaya, A., Levy, N., Scholten, P., & Jansen, J. (2019). Mapping migration studies: An empirical analysis of the coming of age of a research field. *Migration Studies, 8*(3), 455–481.

Silvey, R. (2004). Power, difference and mobility: Feminist advances in migration studies. *Progress in Human Geography, 28*(4), 490–506.

Stewart, A. (2015). Care or work: The tyranny of categories. In *Care, migration and human rights: Law and practice* (pp. 11–26). Routledge.

Szczepaniková, A. (2006). *Migration as gendered and gendering process: A brief overview of the state of art and a suggestion for future directions in migration research*. https://migrationonline.cz/en/e-library/migration-as-gendered-and-gendering-process-a-brief-overview-of-the-state-of-art-and-a-suggestion-for-future-directions-in. Last accessed 10 Feb 2020.

Vlase, I. (2018). Men's migration, adulthood, and the performance of masculinities. In I. Vlase & B. Voicu (Eds.), *Gender, family, and adaptation of migrants in Europe* (pp. 195–225). Palgrave Macmillan.

Wang, C. (2016). Introduction: The 'material turn' in migration studies. *Modern Languages Open*. https://doi.org/10.3828/mlo.v0i0.88

Warren, A. (2016). Crafting masculinities. *Gender, Place and Culture: A Journal of Feminist Geography, 23*(1), 36–54.

Wernesjö, U. (2020). Across the threshold: Negotiations of deservingness among unaccompanied young refugees in Sweden. *Journal of Ethnic and Migration Studies, 46*(2), 389–404.

Wimmer, A., & Glick Schiller, N. (2002). Methodological nationalism and beyond: Nation-state building, migration and the social sciences. *Global Networks, 2*(4), 301–334.

Wimmer, A., & Glick Schiller, N. (2003). Methodological nationalism, the social sciences and the study of migration: An essay in historical epistemology. *International Migration Review, 37*(3), 576–610.

Wojnicka, K., & Pustułka, P. (2019). Research on men, masculinities and migration: Past, present and future. *NORMA, 14*(2), 91–95.

Xiang, B. (2016). Beyond methodological nationalism and epistemological behaviouralism: Drawing illustrations from migrations within and from China. *Population, Space and Place, 22*, 669–680. https://doi.org/10.1002/psp.v22.7

Zapata-Barrero, R. (2019). Methodological interculturalism: Breaking down epistemological barriers around diversity management. *Ethnic and Racial Studies, 42*(3), 346–356.

Zhu, H., & Qian, J. (2020). New theoretical dialogues on migration in China: Introduction to the special issue. *Journal of Ethnic and Migration Studies*. https://doi.org/10.1080/1369183X.2020.1739372

Chapter 3
Gendered Labour

As we saw in Chap. 1, the gendered transfer of labour globally and within Europe has been the focus of attention and the core of the discourse concerning the feminization of migration. Whilst gendered labour migrations are not new, their composition, extent, and how we analyse them, theoretically and methodologically, have evolved. As data show, migrants and especially females, are heavily concentrated within certain sectors producing not just a migrant division of labour (Wills et al., 2010) but a gendered migrant division of labour. Some sectors such as household services (domestic work and care) or social reproductive labour are not only predominantly female but, especially in Southern Europe, overwhelmingly filled by migrant women. Although this type of work has attracted much attention in studies of female labour migration, other sectors, both lesser skilled and more skilled, have also relied heavily on female migrant labour but have been much less studied. Mirjana Morokvasic (2011) questioned the basis of our preoccupation about migrant women as subaltern and victims, exclusively filling low skilled sectors. Thus domestic and care workers have become the emblematic figures of globalised migrations in stark contrast to the easily mobile male IT worker (Kofman, 2013). This is not to deny that domestic and care work globally employ more migrant women than any other sector, and that demand has not grown in response to the inadequacies of public provision across different welfare regimes, leading to the search for cheap solutions to fulfil reproductive needs by using migrant workers, including men. However it does raise issues around our lack of attention to other low skilled sectors such as hospitality and contract and commercial cleaning in hospitals, offices and public spaces, which also employ large numbers of migrants. Skilled labour (IOM/OECD, 2016), especially in welfare sectors, such as education, health and social work is also sourced globally to make good shortfalls in professional reproductive labour (Kofman & Raghuram, 2015). Thus at all skill levels migrant women are employed disproportionately in diverse sectors of social reproduction in sustaining the wellbeing of the household and of society more generally.

Theoretically the focus on domestic and care work and the privileging of the household as the key site of labour (Kofman, 2013) adopted the framework of an

A. Christou, E. Kofman, *Gender and Migration*, IMISCOE Research Series,
https://doi.org/10.1007/978-3-030-91971-9_3

international transfer of gendered labour based on the concept of global chains of care (Hochschild, 2000). Despite its critiques and extension to nursing as a skilled sector (Yeates, 2009), we argue that we should consider a broader analysis of gendered division of migrant labour drawn from a range of sources, European and beyond. We also need to place the demand for migrant labour in a broader context of the transformation of labour markets, deregulation and the intensification and embedding of neo-liberal capitalism, so as to capture the different sectors, skills and sites in which migrants work (Kofman & Raghuram, 2015; McDowell, 2016; Williams, 2018).

Methodologically studies have often focused on a particular sector without tracing the trajectory of individuals into and out of a sector over time (Ryan et al., 2016). Yet the aspirations, personal projects and changing life course shape the attitudes of migrants towards their working lives and strategies they use to deal with their situation. Many remain trapped in precarious, low paid and devalued work but even within low skilled work, there may be opportunities to move out of the most exploitative strata, for example through becoming regularized and accessing pub- licly provided services, as Moré (2019) shows for domestic workers in Barcelona. Those who have a legal status even where there is little exit from the domestic and care sector may find satisfaction in their employment if earning decent wages, highlighted in a quantitative study in Italy (Barbiano di Belgiojoso & Ortensi, 2019).

Others may take entry level jobs, such as domestic work for women and con- struction for men, as they learn the language and how to navigate the labour market before managing to move into better paid and higher level employment, as has been the case of a number of Eastern Europeans in the UK (Parutis, 2014). The ability to resist, find alternative employment and contest individually or collectively (see Chap. 6) their working conditions and discriminatory practices they face will depend on class, gender, race and age as well as on legal and employment status. Although migrants may face exploitation, deskilling and over qualification in their workplaces, emancipation, empowerment and the fulfilment of personal projects may also be the outcome of migratory trajectories.

In this chapter, we firstly trace the development of analyses of gendered migrant labour which tended to focus on female labour migration with less attention paid to male migrants or to predominantly male sectors. Secondly, we outline how the growth of domestic and care work led to theorisations of the global transfer of labour and the focus on this sector in relation to women's increasing participation in the labour market, neo-liberalisation of welfare provision and the commodification of this kind of work. Thirdly, we extend the study of gendered labour to encompass other less skilled and skilled sectors and highlight the stratification between and within sectors. Finally we examine overqualification and deskilling faced by many migrants and suggest this should be understood within a transnational and spatio- temporal perspective.

3.1 Early Studies of Female Labour Migrations

The analysis of women and migration in the 1980s (Morokvasic, 1984; Phizacklea, 1983) sought to make migrant women visible in the labour force and examine how they were incorporated into labour markets. The dominant narrative was of a post war migration comprising males, usually single, with women entering as dependants after the stoppage of labour migration in the early 1970s. This view, however, was contested by a number of scholars. Even though men were the majority, women contributed significantly to labour recruitment as domestic workers, in manufacturing for their nimble fingers, and in the health sector as nurses (Kofman et al., 2000). The flows included women from the Caribbean (Byron & Condon, 2008), Eastern Europe (McDowell, 2005, 2009), Southern Europe (Oso, 2005) and from Turkey (Erdem & Mattes, 2003). From the 1980s female migrants clearly dominated flows in Southern Europe (Campani, 1993) as the need for domestic labour, formerly supplied by internal migration (Escriva, 1997), emerged to fill gaps in provision. By the 1990s, Eastern Europeans could move within Europe without visas and began to contribute to the supply of labour. Female migrants from Poland to Germany, for example, adopted pendular mobility or 'settling in mobility', often moving for short periods between neighbouring countries to work in rotation as cleaners, babysitters and care workers or traders (Morokvasic, 2004). As interest turned to international labour migration to Europe and North America, the earlier concern with internal migration in the Global South (Bunster & Chaney, 1985; Chant, 1992) slipped into the background.

At the beginning of this century two key scholars (Morokvasic, 2007; Phizacklea, 2000) reflected on what had changed in the past two decades of feminist research on migrant women and labour markets, what lessons could be drawn from this body of work, and how it could be situated in relation to broader theoretical developments in migration and economic and political changes in Europe and globally. In reviewing 30 years of studies of women migrant workers, Morokvasic reflected that gender orders and hierarchies have not been overturned. The segregation of women into a number of generally low paid reproductive occupations has reinforced gender hierarchies. For Morokvasic (2007: 92) the reproduction of the gender order in migration reveals contradictory outcomes. She argued that most migrant women look for 'compromises rather than confrontation and rejection of the gender division of labour and values'. Differences in terms of class, ethnicity, age and occupation are likely to shape such outcomes. In a recent study of Ukrainian care givers of elderly persons in Italy, Tyldum (2015) concludes that exploitation and empowerment may coexist. Though the conditions of work may be exploitative in Italy, it is often better than the situation in the Ukraine where they are often performing similar care work unpaid and dealing with problematic marital relationships with limited access to divorce.

Similarly, looking back at several decades of studies of female migrant workers, Phizacklea (2000) critiqued the earlier structural neo-Marxist political economy analyses treating migrant women as cogs in the capitalist world order. She advocated endowing migrants with more agency through a structure-agency approach (Goss &

Lindquist, 1995) and engagement in transnational processes from below as a means of subverting the logic of transnational capital. At the same time, she noted that this may not be equally available to all, as we see in the access to the internet and social media.

The view of the role of migrant women in the labour force since the earlier waves of migration has also changed. In the 1970s domestic work was considered to be pre-modern, evoking labour contracts and relationships of bygone eras (Friese, 1995), and thought likely to die out. What demand existed was being provided by working class women and/or internal migrants, although a few studies highlighted the presence of women from the European periphery in the wealthy areas of cities, as with Spanish maids in Paris (Oso, 2005). Nonetheless, some writers (Gorz, 1988) noted the increase in the servant class in the 1980s in response to the push towards economic rationality and growing inequalities in income between the middle classes and working classes whose work was far less well remunerated.

3.2   Care and Social Reproduction

It was the emergence in the 1990s of domestic and care work as forms of reproductive labour across a range of welfare regimes in the Global North, which generated extensive empirical research and a theorization based on the transfer of labour within a global system. Care can be defined "as the work of looking after the physical, psychological, emotional and developmental needs of one or more other people" (Standing, 2001: 17) which includes a wide range of people requiring care, some of whom are vulnerable (children in care, homeless, frail elderly, mental health) or with a pronounced degree of dependency (young children, those with disabilities). The commodification and marketisation of care, especially after the financial crisis of 2008 and ensuing austerity policies, led to downward pressure on the remuneration of domestic and care workers. The number of migrants has expanded dramatically to meet the inadequacies of welfare provision and ageing populations (Farris, 2015; Lutz, 2008). The female domestic's work, though increasingly indispensable, is lowly valued, poorly remunerated and minimally recognised in European immigration policies. The skills required for these services are embodied (McDowell, 2015; Wolkowitz, 2006), transferred from practices in the domestic sphere, and thereby depicted as innately female.

Theoretically, the concept of global chains of care (Hochschild, 2000; Parreñas, 2001), denoting the transfer of emotional and physical labour from poorer households in the Global South to those in the Global North, became the dominant lens through which this transfer was construed. In turn, those in poorer countries generated another chain, usually drawing in poorer women or other family members to supply the care deficit, though this aspect only drew attention some time later through an interest in children and elderly left behind (see Chap. 4 on transnational families). The concept caught on rapidly and was quickly applied to the use of migrant labour from beyond Europe and later within Europe from poorer countries in the East and South East as well as on Europe's borders (Lutz & Pallenga-

Mollenbeck, 2012; Palenga-Möllenbeck, 2013; Vattinen, 2014). Indeed in Europe, we can discern cascading care chains with migration from Eastern and South Eastern Europe to Western and Southern Europe and the replacement of labour by migrant women from beyond the EU, for example, Ukrainians to Poland and the Czech Republic (Sowa-Kofta, 2017).

However, the concept has been subject to a number of criticisms. In its initial version, the global chains of care focused narrowly on transnational mothers as the vehicle for transferring care, thereby ignoring the familial diversity of transnational carers and its heteronormative assumptions (Manalansan, 2006). It has also been critiqued for its narrow focus on a particular form of care, failure to take into account changes in caring relationships and circumstances throughout the life course.

While child care underpinned the global chains of care perspective, care for the elderly has come to play a much bigger part and the elderly person, female or male, may themselves be the direct employer of the worker (Cangiano et al., 2009). So too have elderly family members, often unpaid, formed a care chain quite different to the original version. Transfers of labour also included older women, as in the case of Ukrainian women to Italy (Tyldum, 2015) and migrant men supplying care, especially for the elderly (Gallo & Scrinzi, 2016).

More recently, new perspectives have sought to place recourse to migrant labour in a broader context of the political economy of care and transnational labour operating at different scales (micro, meso and macro) (Williams, 2018) and across different sites of the market, state, NGOs and households in what has been called the care diamond (Razavi, 2007; Kofman & Raghuram, 2015). The crisis and contradictions of care in neo-liberal capitalism, especially after the 2008 financial crisis (Fraser, 2016) and the role of the state through its employment and social policies in shaping household demand for services (Carbonnier & Morel, 2015; Shire, 2015) are also seen as having contributed to marketisation and privatisation of low waged and low skilled labour (Aulenbacher et al., 2018). The discourses behind the expansion of household services in a number of European states (Austria, Belgium, France, Germany, Sweden) have revolved around providing employment for the low skilled and curbing informal employment, responding in a more cost effective way to social needs through private schemes, promoting female employment and supporting the productive potential of the highly skilled. Whilst care activities especially for the elderly and disabled, have been targeted through the use of vouchers, tax credits and legal employment of 24 hour carers, as in Austria and Germany (Haubner, 2020), eligible household tasks encompass a range of services well beyond care, including cleaning, home maintenance and gardening in other countries (France, Sweden).

This range of activities serves to support the social reproduction of households, especially those of high income households who tend to be the main beneficiaries. By social reproduction we mean the production of people through various kinds of work – mental, manual and emotional – aimed at providing what is necessary to maintain existing life and to reproduce the next generation. It includes:

> how food, clothing and shelter are made available for immediate consumption, the ways in which the care and socialization of children are provided, the care of the infirm and the elderly, and the social organization of sexuality...And the organization of social reproduction refers to the varying institutions within which this work is performed, the varying

strategies for accomplishing these tasks, and the varying ideologies that both shape and are shaped by them (Laslett & Brenner, 1989).

Social reproduction thus encompasses a range of activities, including but not restricted to care, within and beyond the household. And very importantly it is not limited only to those who are dependent such as children, the frail elderly or the disabled. It also includes tasks often undertaken by migrant men such as household maintenance (Kilkey et al., 2013), gardening (Ramirez, 2011) as well as domestic work and care (Davalos, 2020; Gallo & Scrinzi, 2016).

Given that much of the employment generated by these employment-generation policies has been part-time and of variable hours, low skilled and poorly valued, that is non-standard employment relations, it is not surprising that migrants, and especially recent as opposed to more established ones, have disproportionately filled these sectors. Furthermore, income earned from mini jobs in Germany is capped (Haubner, 2020) whilst in Belgium and France travel time is not included (Morel & Carbonnier, 2015).

The role of migrants in providing household labour differs between countries and regions, types of tasks involved and whether live-in or live-out. It depends on immigration policies, welfare regimes, gender ideologies and cultural attitudes towards provision of care within the household and externally. The countries with the highest reliance on migrant labour for domestic work and care are the Mediterranean ones which are part of familial welfare regimes in which migrant women in particular have replaced the housewife and other family members in household tasks and care of children and the elderly. In Italy, for example, 90% of domestic workers were foreign-born (Rostgaard et al., 2011). In Eastern Europe, on the other hand, the employment of migrant labour in the household is well below the European average. The use of migrants in the household also varies between tasks eg. cleaning and care for children and the elderly (Box 3.1).

Box 3.1: Outlines the Hierarchy Between Different Forms of Household Employment in France

Within household employment, there exists a hierarchy. For example, the foreign-born in France comprised 40% of cleaners but only 15% of child and elder carers (Avril & Cartier, 2014). Among domestic workers overall, 60% had no qualifications with 29% of the group being foreign; among home caregivers 36% had no educational qualifications with 8% of this group being foreign-born according to the 2011 French Employment Survey (Devetter & Lefebvre, 2015: 163). Cleaning, whether domestic or contract, is often an entry job, especially for those without knowledge of the language (Abbasian & Hellgren, 2012; McIlwaine, 2020; Parutis, 2014). In the Swedish study (Abbasian & Hellgren, 2012: 171), there were, on the other hand, over 20% female and male migrants with a university degree. Although there may be an overlap between domestic work and care givers, the symbolic significance and social recognition of the two is quite different (Devetter & Lefebvre, 2015: 163).

The extent to which migrants are used in household work also varies between localities and regions. It is likely that migrants contribute to household labour more in large cities. In the UK, while London and the South-East have high percentages of migrants both from the EU and non-EU, in most other regions, the percentage is much lower. In London only 60% of care workers are British compared to the North East, the region with the lowest percentage, with 96% British. Similarly the proportion of minority ethnic groups employed in the care sector is much higher in London where some local authorities have over 80% from minority ethnic groups in care employment (Howard & Kofman, 2020).

3.3   Understudied Sectors and Gendered Migrant Division of Labour

Though domestic and care work have provided a major source of employment globally (ILO, 2018), other forms of labour have also generated employment with non-standard employment relationships, such as part-time, fixed term, zero hours, and precarious work (Vosko, 2010) for migrant workers as states strive to fill low paid and often poorly regulated work. In the EU, both EU migrants from Central and Eastern Europe as well as non-EU from Latin America, Africa and South East Asia provide labour across the different sectors. We should also note that those in the labour force may enter through a variety of routes (van Hooren, 2012), such as family reunification (see Chap. 4), asylum seekers and refugees (see Chap. 5) and students (Maury, 2020). So, while construction is overwhelmingly male dominated, and household services female, in OECD countries the majority of sectors are to varying degrees mixed as Table 3.1 shows.

For EU-27 countries (see Table 3.2), migrant labour is also spread across a number of sectors beyond those employed by households. It is only in Southern Europe, in what has been called the migrant in the family welfare model (Bettio et al., 2006), that migrant women are concentrated in the household as employer

Table 3.1 Gendered division of migrant labour by sector (percentage)

Sector	Male	Female
Agriculture and fishing	71.9	28.1
Manufacturing, mining, energy	67.8	32.7
Construction	93.8	6.2
Wholesale and retail trade	57.8	42.2
Hotels and restaurants	55.4	44.6
Education	37.7	62.3
Health and social work	40.6	59.4
Household services	15.4	84.6
Administration	53.6	43.7
Other services	59.5	40.5

Source: OECD (2017)

Table 3.2 Employment of male and female native- and foreign-born workers, 25–54 years, by occupational category in EU-27 in 2008[a]

Sector	Native-born	Foreign-born
Men		
Manufacturing	22	22
Construction	13	19
Wholesale and retail activities	13	12
Accommodation and food	2	8
Transportation and storage	8	8
Administration and support services	3	5
Human health and social work	4	4
Professional, scientific, technical	5	4
Information and communications	4	3
Public administration, defence, security	8	3
Women		
Human health and social work	17	18
Wholesale and retail	15	13
Manufacturing	12	11
Accommodation and food	4	10
Activities household as employers	1	10
Administration and support services	4	8
Education	4	4
Professional, scientific, technical	5	4
Other services	3	4
Public administration, defense	8	4

Source: Eurostat (2011: 46)
[a]The subsequent 2014 ELFS ad hoc module focuses primarily on indicators of labour market integration, such as activity rates, employment status, unemployment and overqualification rather than sectors

sector. Such a concentration is much greater than in any of the male dominated sectors such as construction. In the accommodation and food sector, both foreign-born women and men are disproportionately present. However, the important health and social work sector, covering a variety of skill levels and sites of public and private sector employment, is also one in which there is generally not a large disparity between the presence of native and foreign-born proportion of the working population though they may occupy different strata within an occupation.

The almost exclusive focus on the female domestic/care worker in feminist research of gendered labour markets, though understandable, is problematic. It reinforces stereotypes of migrant women (Catarino & Morokvasic, 2013) and fails to recognise the much broader gendered migrant division of labour extending across a diversity of skill levels and sites (private and public). As we have outlined, the household has captured our attention and become a privileged sector of employment for the less skilled encouraged by state and EU policies. However as both Michael Bittman et al. (1999) and Mignon Duffy (2005) have asked, though in slightly different terms, why do we not study the outsourcing of labour from the household stemming from changing consumption patterns. Sassen's (1992) dissection of the global city has also drawn attention to this aspect. Precarious employment typifies both the labour insourced (domestic and care work, gardening) into the household as well as that which is outsourced, such as production of ready-made food and its external consumption, which have dramatically altered reproductive labour in the household. This outsourcing has been accentuated in large and global cities, resulting in precarious employment becoming a dominant feature of social relations between employers and workers in the contemporary world (Standing, 2011) and constitutive of a new global disorder (Schierup et al., 2014). Welfare restructuring has also led to costs of social reproduction and transactional costs (making applications, travel to interviews for a series of temporary employment) of entering and continuing in the labour market being increasingly borne by individual and families. Thus the growth of global labour migrations has been accompanied by the intensification of non-standard employment relationships, contracting out of services and deregulation of labour. And as Linda McDowell (2009: 7) has commented there is a "hierarchy of desirability within the category of 'economic migrants'".

Among the less skilled, the gendered division of labour encompasses a wide range of sectors (Amrith & Saharoui, 2019; Dyer et al., 2011). Female-dominated sectors or those with substantial numbers beyond those employed by households, include hospitality (Adib & Guerrier, 2003; Batnitzky & McDowell, 2013; McDowell et al., 2009, 2012), contract cleaning in hospitals (Stournara, 2020; von Bose, 2019), offices and public spaces and bodywork, such as beauty parlours, hairdressers, manicurists (McDowell, 2009; Wolkowitz, 2006) and sex work, to name a few. What is seen as dirty work is often divided along gender lines such that cleaning and care are undertaken by women while refuse collection and street cleaning are male domains. Work that is physically tainted or physically dirty may be identified with particular class, race and migrancy characteristics where workers' bodies are seen as being suitable for this kind of work (McDowell, 2009; Wolkowitz, 2006) or what has been called 'suitable embodiment' (Simpson & Simpson, 2018). Privatization and worsening and precarious working conditions and social security since the 1990s have made these jobs even more 'suitable' for migrants than ever. For example, in Sweden at the beginning of the century cleaners represented one of the 20 most common occupational groups with 80% of women and 31% of foreign origin, of whom those in hourly and part-time work were more likely to be from non-EU countries. Among migrants there tend to be more men in this sector than among the native population.

Sex work has been commonly associated with trafficking and the sexual exploitation of women (Kempadoo, 2005) without recognising that it may stem from other forms of work or that women, men, transgender people and children can be trafficked into diverse forms of exploitative labour (Howard, 2019). Nonetheless the gendered and sexualised victimisation of migrant women is still a dominant paradigm in the field (Palumbo & Scuirba, 2018). Sex work has also generated polarised views. So whilst some have portrayed women as passive victims of trafficking, a number of authors have begun to challenge this particular framing which presents women in a way, which negates their agency and decision-making capacity when opting for sex work (Agustin, 2005). 'Sexual humanitarianism' (Mai et al., 2021) is a concept which analyses how migrant sex workers (men and women) are impacted by policymaking and social interventions based on their presumed vulnerability to trafficking and exploitation (see Chap. 5) and "neo-abolitionist discourse, which systematically conflates prostitution with trafficking, seeking its abolition by removing the demand for sexual services".

Studies using ethnographic methods, including films, have given women their voice, showing how they decided to undertake sex work. Research over a number of years with women from the North of China who had migrated to Paris after losing their jobs following economic restructuring in China in the 1990s. (Lévy & Lieber, 2009) showed how having found themselves in an unexpectedly precarious situation, they utilised sex as a resource through diverse sexual-economic arrangements, such as cohabitation with a fellow national, marriage with a French man which could lead to regularisation of their legal status or prostitution. The latter situation, though looked down upon, brought them much more money than working for South Chinese families as child minders which was the most demeaning for them. Nigerian women in Italy have been associated with sex work and therefore assumed to have been trafficked (Plambech, 2017), which may make it difficult for them to have their asylum claims or the violence they experienced in their home countries and on their journeys, taken seriously (Rigo, 2017). In Italy and Spain many migrant women, including from the recent EU enlargement countries such as Romania, are hired for seasonal and temporary labour in agriculture. Their dependence for future contracts may make them vulnerable to exploitative sexual relations (Palumbo & Sciurba, 2018).

The dominant focus on less-skilled employment pushes into the background the circulation of skilled female migrants and endorses the paradigmatic separation of (male) skilled and (female) less-skilled understandings of migration. Females working in skilled sectors tend to dominate reproductive sectors, such as health, social work and teaching, which often do not pay as much as male occupations, although many educated women work in mixed or male dominated sectors (Raghuram, 2008). So, too, is gender largely absent in studies of international student migration (Sondhi & King, 2017; Raghuram & Sondhi, 2021) due in part to the assumption that female migrants are largely of working class origin, yet the gendered mobilities of students may feed into flows of skilled migrants.

3.4   Skilled Sectors and Gendered Migrant Division of Labour

Skilled immigration, as Boucher (2007) commented "has slipped by as a genderless story in which the androgynous skilled migrant is the central character and economists do most of the storytelling". Meares (2010: 473) stated that "despite the now significant body of scholarship on the relationship between gender and international migration, scant attention has been paid to the gendered transition experiences of highly-skilled migrants". Yeoh and Willis (2005) too make a strong argument for more research in this area, noting that existing work on the devalued and often "racialised" labour of unskilled women must be complemented by a greater focus on professional and entrepreneurial women who remain largely absent in the broader analysis of 'transnational elites'. Yet the current global race for talent in which states seek to attract skilled migration, "is profoundly gendered, with significant implications for the skill accreditation, labour market outcomes, rights of stay, and gendered family dynamics" (Boucher, 2016: 30).

Furthermore, Dumitru (2017) argues that theories of feminization linked with the analysis of globalization and international division of labour have tended to ignore the educational qualifications of women. Dumitru asks whether we should be referring to feminization of migration (see Chap. 1) as skilled migration since women migrants are increasingly educated and a higher percentage of those with tertiary education are migrating more than men (Docquier et al., 2009; Dumitru, 2014). For example, tertiary educated among female migrants rose from 18% in 1980 to 40% in 2010 so that women now form a majority of skilled migrants (Weinar & Klekowski von Kopenfels, 2020). Discriminatory practices in employment, family practices and public participation may also act as drivers of migration (Ruyssen & Salamone, 2018). However being highly educated does not mean either entering through skilled routes or working in skilled occupations (Boucher, 2020; Carangio et al., 2021), hence the conversion of education into skilled employment post migration may be fraught with barriers, especially where income constitutes a major determinant of entry through skilled channels (Boucher, 2016; Kofman, 2014), as in the UK and the EU Blue Card. Women in skilled sectors are concentrated in regulated professions where they encounter barriers to recognition and consequently circulation.

Furthermore, large numbers of educated female migrants enter through family migration (see Chap. 4) but we know little of their work aspirations or experiences. There is some evidence that they may face particularly high levels of deskilling (Ballarino & Panichella, 2018; Triandafyllidou & Isaakyan, 2016 (see Sect. 3.5). Migrants, and particularly the skilled, who are able to benefit from bringing an accompanying spouse or being reunified with them, will frequently migrate together. For example, among Indian migrants many males enter the IT sector, while female highly educated spouses often come with or follow them shortly afterwards and succeed in entering the labour market (Kõu et al., 2015). Though entering the labour market, they are often simply described as trailing spouses (van Bochove & Engbersen, 2015), a term that goes back to the 1980s when there was a concern about obstacles to the international migration of those in management positions (Weinar & Klekowski von Koppenfels, 2020). It is only with more interest in the gender dimensions of skilled migration that questions are being asked about spouses of the highly skilled who themselves are highly educated.

In general, studies of skilled migration streams focus on the economic realm where the social dimension barely intrudes. Studies of skilled migration reproduce the notion of economic man and social woman (Kofman & Raghuram, 2005; Schaer et al., 2017). As Kõu et al. (2015) state: "There has been little attempt to link highly skilled migrants with life course analysis so as to study their parallel careers of migration, employment and household... and gain a deeper understanding of the influence of life paths, social networks, diasporas and immigration policies".

> **Box: Gender Inequalities, Academia and the Lifecourse**
> One sector in which studies have explored how gender inequalities are reproduced through work demands and life course changes is that of young academics and researchers (Ackers & Bryony, 2008). Studies (França & Padilla, 2017) have found that academic careers follow a male linear career structure with little room for family responsibilities to impinge and thereby reproduce gendered inequalities, such as glass ceilings, gender pay gap, sexual harassment and exclusionary dynamics, found in other domains of the labour market However, Schaer et al. (2017) note that gender inequalities in academic mobility may acquire a complexity beyond a traditional representation. They reflect on examples of both female and male tied movers to counteract a simplistic view of the trailing spouse and a rather negative view of what spouses do when they accompany partners. In some cases, and especially when it is planned, the spouse may elect to train or to study (Raghuram, 2004). Yet in other cases, women's migration may lead to downward mobility for their husbands, who find themselves doing jobs as security guards, retail and carers, as was the case of spouses of Nepalese nurses in the UK (Adkikari, 2013).

Studies of health care professionals have tended to focus on macro issues of supply and demand with much less attention paid to working conditions or the family and social dynamics of their lives. It is a sector where, despite reductions following the 2008 financial crisis, recourse to migrant labour subsequently increased to fill shortages due to an increase in demand as a result of population ageing, technological advances and higher patient expectations (Castagnone & Salis, 2015), combined with inadequate supply arising from attempts to cut back expenditure and training as well as poor working conditions (Yeates, 2010). Nursing, as a mobile profession, has become a major sector of migrant women's employment globally (Kingma, 2006) based on an export-oriented production system combined with recruitment strategies from wealthy countries (Wojczewski et al., 2015; Yeates, 2009). In Europe, those from new EU countries in particular have filled the gap. In most countries, with the exception of Germany, there has been a reliance on recruitment agents to navigate entry restrictions and barriers to recognition of credentials which are faced by non-EU healthcare professionals in Europe and in other major countries of shortages, such as Australia, Canada and the United States (Walton-Roberts, 2021a). Together with a range of actors, including the state and recruitment agents in both countries, connections between countries form complex global and highly dynamic chains of nurse care framework (Yeates, 2009) and

Table 3.3 Percentage of female migrants in the healthcare workforce

Sector		Germany	Spain	Italy	UK
Doctors and health professionals	Foreign	72.3	47.0	64.4	57.9
	National	72.1	58.2	49.6	61.1
Nurses	Foreign	90.1	89.9	83.1	82.5
	Nationals	86.0	83.3	75.5	90.2

Source: Villosio (2015)

pathways (Walton-Roberts, 2021b), encompassing both circuits in the Global North and between the South and the North. Thus the United States takes nurses from Canada which recruits both from the UK and countries in the South. The UK in turn is the main recruiter in Europe attracting both those from the European periphery, especially Portugal and Spain after the 2008 crisis and Romania since enlargement, as well as countries in the South. Certain countries such as the Philippines, India and Vietnam have developed nurse export strategies and emerged as the main source countries for nurse migration (Thompson & Walton-Roberts, 2019; Yeates, 2010).

The pathways from poorer countries to their insertion in the wealthier are not always smooth for nurses (Näre & Nordberg, 2016). They may confront racialized attitudes from other healthcare workers as well as patients. Data on their working conditions in selected European countries reveal that they may have to do more unsocial shifts and have less secure contracts (Castagnone & Salis, 2015).

Although nursing is an overwhelmingly feminised profession (see Table 3.3), there are male migrants who in the UK are more prevalent than native men. Some come from countries where men have trained as nurses in the country of origin as with the Philippines, while others come from countries where it is unusual for men to work in this sector but who have taken it up as a profession at a time when there were pronounced shortages and opportunities for training. This was the case with Zimbabwean men who as black men also encounter racism at work. Yet as Panopio (2010: 12) comments: 'there is a lack of contextual analysis that allows for inter-sections of race, class, sexuality, and other identity formations in mainstream accounts of male nurses and international nurse migration'. In Panopio's study of Filipino male nurses in London, most had been helped by family members into a profession that endowed them with social mobility and would enable them to help others in turn. Some, especially gay men, also wanted a softer or more feminine environment to work in but within nursing tended to orient themselves to male spaces of work, as in operation theatres. And the possibilities of social mobility through a global profession attenuated to some extent the feminine resonances of the profession.

In contrast, there is little gendered analysis of masculine sectors of work (Grigoleit-Richter, 2017; Raghuram, 2008), such as IT, which unlike the regulated professions have little state control of accreditation and for which barriers to mobility are lower, especially for those only moving temporarily. As Raghuram (2008) comments, it is particularly interesting to compare female and male migrants in this sector. In Europe and other major countries of permanent migration, the number of women in STEM has declined while in a number of sending countries of highly skilled ICT migrants, such as India, women form a notable proportion of students in STEM subjects (over 40%). So too has ICT employment grown rapidly among women in India (30% in 2016) compared to its stagnation in many countries in the West. This is not a migration of survival; for women in particular, it gives them the opportunity for career development, travel and, for some, to move away from restrictive cultural and social practices (Kõu et al., 2017). The intersectionality of gender and class plays out very differently to other skilled sectors such as nursing which is at the lower end of health care professions. Indian ICT migrants, for example, are from a solidly middle class background and women, even more than men, have both parents who work and mothers with tertiary degrees (Sondhi et al., 2018).

3.5   Deskilling and Devaluation

Despite the growing number of highly educated migrants (Dumitru, 2017; OECD, 2018), both those from the European Union and non-Europeans have faced considerable levels of deskilling (Sert, 2016, OECD, 2018) and devaluation of their cultural capitals (Bourdieu, 1986). Over-qualification (higher educational qualification than what is required for the job) in Europe affects 36% of migrant women and 31% of migrant men compared to 22% of native women and 20% of men. It is particularly pronounced in Southern Europe countries where for migrant women it may reach 50% level of over-qualification (OECD, 2018).

The poor labour market integration of highly educated migrant women is linked to issues of foreign degree recognition, emphasis on host country work experience and a preference for local accents in relation to language skills. The latter is shown to disadvantage women, particularly considering their concentration in relational work such as support, service, and caring labour, and in regulated sectors such as health care, in contrast to male-dominated financial and technical occupations. National privileges may result in occupations being reserved for national and EU citizens forcing migrants to be employed on less secure and less well paid contracts, for example, with doctors and nurses in hospitals (Castagnone & Salis, 2015). Differences in national regulations and difficulties of obtaining accreditation may mean that migrant nurses are unable to continue to undertake routine duties in the destination country and are forced to work instead as nursing assistants or care workers in residential homes and private households (Cuban, 2013).

Racial discrimination on the part of recruiters and co-workers may mean that they are forced to accept positions they are overqualified for, or do not have the same opportunities for career progression as co-workers (Wojczewski et al., 2015). Female migrant workers therefore often face a double penalty in terms of labour market segregation and discrimination. And as previously discussed in relation to dirty work, migrant women and men, such as Eastern Europeans (Fox et al., 2012) and Latin Americans (Cederberg, 2017), may be racialised and stereotyped as being appropriate to perform certain kinds of low skilled work. Women from Central and

Eastern Europe in the UK, as Samaluk (2016) comments, are portrayed either as objects of desire for front-line service work or as traditional and docile workers suitable for domestic and care work. Indeed Favell (2008: 711) presented the dangerous outcome of free movement of "ambitious 'New Europeans' … becoming a new Victorian servant class for a West European aristocracy of creative-class professionals and university educated working mums," with female migrants holding a teacher's diploma or even a PhD working in Austria in the fields of child or geriatric care.

Migrants may thus experience contradictory class position (Parrenas, 2000) in the course of migration, consisting of a lower status job but earning more money in the destination country than in their homeland. Here transnational contexts (Nowicka, 2013, 2014) and family class background play a major role in the extent to which their institutionalised cultural capital or educational qualifications are transposed. As Oso (2020) highlights for Spanish migrants to Paris, even those with degrees find it more difficult to get a job commensurate with their qualifications if they come from a family of modest means who could not assist them or did not have social networks or social capital.

Migration researchers (Cederberg, 2017; Erel, 2010; Kelly, 2012; Nowicka, 2013, 2014; Oliver & O'Reilly, 2010; Ryan et al., 2015; Samaluk, 2016) have increasingly turned to a Bourdieusian analysis (Bourdieu, 1986) to gain a better understanding of the modalities of the international transfer of cultural and other capitals between different societies. As we have seen, linguistic capital (ability to communicate as well as accents) is essential in accessing skilled sector jobs and moving out of elementary employment that may not require much language proficiency. For many migrants, this entraps them in menial work whilst others are able to improve their linguistic capital and move after a time into higher level employment as with Polish migrants in the UK (Ryan et al., 2016). An analysis of how Central and Eastern Europeans found their jobs in Western Europe indicates that a much higher proportion than natives used social networks and there was a correlation with overqualification. Hence their social capital is likely to have facilitated obtaining a job but at the cost of being in a job below their educational levels. The link with linguistic proficiency is less clear as in fact those in hospitality where many find themselves despite having a reasonable linguistic level (Leschke & Weiss, 2020). Attitudes to the disparity in status may differ according to their migratory projects and the extent to which this is aligned with social mobility and status in the country of origin.

3.6 Conclusion

As we have outlined in this chapter, a gendered division of labour encompasses a broad range of both less skilled and skilled sectors. Migrant women are concentrated in feminised sectors of social reproduction ranging from domestic work, care and nursing. Migrant men too are more likely to be found in feminized sectors and, as with women, often deskilled. However, over-qualification and devaluation of migrant cultural capitals resulting in positions of contradictory class mobility need to be situated within a transnational lens, take account of the migratory projects pursued and the intersectionality of class, gender, race and nationality. For some deskilling is a temporary situation either in relation to social mobility within the country of destination or their social status in the country of origin; for others it represents a long-term entrapment. Loss of employment may also lead to onward migration for some through free movement in the European Union (Ortensi & Barbiano di Belgiojoso, 2018) and to complex mobility pathways (Parreñas, 2020).

And at the time of completing the book, the major upheaval has been the global pandemic with countries locking down and closing borders to migration, thus disrupting migratory flows, especially of those circulating regularly, as with some care workers between Poland and Germany or Romania and Austria and seasonal agricultural workers. At the same time, the public came to realise the value of workers categorised as low skilled and of little worth but now deemed as essential and key workers (Fasani & Mazza, 2020; Rasnaca, 2020). On average in the EU, migrants (EU and non-EU) represented 13% of key workers but in some sectors considerably more, especially among non-EU migrants as in Cyprus, Germany, Italy and Sweden. In particular two groups stand out: agricultural workers who guarantee the food chain operates smoothly and ensure food security and care workers, enabling our physical and mental well-being. The evidence shows how much we rely on migrant workers to provide labour in the key occupations, including the supposedly low skilled ones.

During the pandemic itself a number of countries regularised the undocumented in key employment sectors and most significantly there have been calls to acknowledge the need to reconsider immigration policies prioritising the highly skilled and devaluing the less skilled (Fasani & Mazza, 2020). It is not clear whether the essential work that migrant women and men undertake will continue to be appreciated, although it does seem clear that the work they perform will persist despite higher levels of unemployment.

References

Abbasian, S., & Hellgren, C. (2012). Working conditions for female and immigrant cleaners in Stockholm County – An intersectional approach. *Nordic Journal of Working Life Studies, 2*(5), 161–181.

Ackers, L., & Bryony, G. (2008). *"Moving people and" knowledge: Scientific mobility in an enlarging European Union*. E. Elgar.

Adhikari, R. (2013). Empowered wives and frustrated husbands: Nursing, gender and migrant Nepali in the UK. *International Migration, 51*(6), 168–179.

Adib, A., & Guerrier, Y. (2003). The interlocking of gender with nationality, race, ethnicity and class: The narratives of women in hotel work. *Gender, Work and Organization, 10*, 413–432.

Agustin, L. (2005). Migrants in the mistresses house: Other voices in the 'trafficking' debate. *Social Politics: International Studies in Gender, State and Society, 12*(12), 96–117.

Amrith, M., & Saharoui, N. (Eds.). (2019). *Gender, work and migration. Agency in gendered labour settings*. Routledge.

Aulenbacher, B., Décieux, F., & Riegraf, B. (2018). The economic shift and beyond: Care as a contested terrain in contemporary capitalism. *Current Sociology, 66*(4), 517–530.

Avril, C., & Cartier, M. (2014). Subordination in home service jobs comparing providers of home-based child care, elder care, and cleaning in France. *Gender and Society, 28*(4), 609–630.

Ballarino, G., & Panichella, N. (2018). The occupational integration of migrant women in Western European labour markets. *Acta Sociologica, 61*(2), 26–42.

Barbiano di Belgiojoso, E., & Ortensi, L. (2019). Satisfied after all? Working trajectories and job satisfaction of foreign-born female domestic and care workers in Italy. *Journal of Ethnic and Migration Studies, 45*(13), 2527–2550.

Batnitzky, A., & McDowell, L. (2013). The emergence of an 'ethnic economy'? The spatial relationships of migrant workers in London's health and hospitality sectors. *Ethnic and Racial Studies, 36*, 1997–2015.

Bettio, F., Simonazzi, A., & Villa, P. (2006). Change in care regimes and female migration and public policy. *Journal of European Social Policy, 16*(3), 271–285.

Bittman, M., Matheson, G., & Meagher, G. (1999). The changing boundary between home and market: Australian trends in outsourcing domestic labour. *Work, Employment and Society, 13*(2), 249–273.

Boucher, A. (2007). Skill, migration and gender in Australia and Canada. The case of gender-based analysis. *Australian Journal of Political Science, 42*(3), 383–401.

Boucher, A. (2016). *Gender, migration and the global race for talent*. Manchester University Press.

Boucher, A. (2020). How 'skill' definition affects the diversity of skilled immigration policies. *Journal of Ethnic and Migration Studies, 46*(12), 2533–2550.

Bourdieu, P. (1986). The forms of capital. In J. Richardson (Ed.), *Handbook of theory and research for the sociology of education* (pp. 241–258). Greenwood Press.

Bunster, X., & Chaney, E. (1985). *Sellers and servants: Working women in Lima, Peru*. Praeger.

Byron, M., & Condon, S. (2008). *Migration in comparative perspective: Caribbean communities in Britain and France*. Routledge.

Campani, G. (1993). Immigration and racism in southern European: The Italian case. *Ethnic and Racial Studies, 16*(3), 507–535.

Cangiano, A., Shutes, I., Spencer, S., & Leeson, G. (2009). *Migrant care workers in ageing societies: Research funding in the UK*. COMPAS.

Carangio, V., Farquharson, K., Bertone, S., & Rajendran, D. (2021). Racism and white privilege: Highly skilled immigrant women workers in Australia. *Ethnic and Racial Studies, 44*(1), 77–96. https://doi.org/10.1080/01419870.2020.1722195

Carbonnier, C., & Morel, N. (Eds.). (2015). *The political economy of household services in Europe*. Palgrave Macmillan.

Castagnone, E., & Salis, E. (2015). *Workplace integration of migrant health workers in Europe. Comparative report on five European Countries*. FIERI.

Catarino, C., & Morokvasic, M. (2013). Women, gender, transnational migrations and mobility: Focus on research in France. In L. Oso & N. Ribas-Mateos (Eds.), *The international handbook on gender, migration and transnationalism* (pp. 246–267). Edward Elgar.

Cuban, S. (2013). *Deskilling migrant women in the global care industry*. Palgrave Macmillan.

Cederberg, M. (2017). Social class and international migration: Female migrants' narratives of social mobility and social status. *Migration Studies, 5*(2), 149–167.

Chant, S. (Ed.). (1992). *Gender and migration in developing countries*. Belhaven.

Davalos, C. (2020). Localizing masculinities in the global chains of care: Experiences of migrant men in Spain and Ecuador. *Gender, Place and Culture, 27*(2), 1703–1722.

Devetter, F.-X., & Lefebvre, M. (2015). Employment quality in the sector of personal and household services: Status and impact of public policies in France. In C. Carbonnier & N. Morel (Eds.), *The political economy of household services in Europe* (pp. 150–171). Palgrave Macmillan.

Docquier, F., Lowell, B. L., & Marfouk, A. (2009). A gendered assessment of highly skilled emigration. *Population and Development Review, 35*(2), 297–322.

Duffy, M. (2005). Reproducing labor inequalities. Challenges for feminists conceptualizing care at the intersection of gender, race and class. *Gender and Society, 19*(1), 66–82.

Dumitru, S. (2014). From 'brain drain' to 'care drain': Women's labor migration and methodological sexism. *Women's Studies International Forum, 47*, 203–212.

Dumitru, S. (2017). Feminisation de la migration qualifiée: les raisons d'une invisibilité. *Hommes & Migrations, 1317*, 146–153.

Dyer, S., McDowell, L., & Batnitzky, A. (2011). Migrant work, precarious work–life balance: What the experiences of migrant workers in the service sector in Greater London tell us about the adult worker model. *Gender Place and Culture, 18*, 685–700.

Erdem, E., & Mattes, M. (2003). Gendered patterns: Female labour migration from Turkey to Germany from the 1960s to the 1990s. In R. Ohlinger, K. Schonwalder, & T. Triadafilopoulos (Eds.), *European encounters. Migrants, migration and European societies since 1945* (pp. 167–185). Ashgate.

Erel, U. (2010). Migrating cultural capital: Bourdieu in migration studies. *Sociology, 44*(4), 642–660.

Escriva, A. (1997). Control, composition and character of new migration to south-west Europe: The case of Peruvian women in Barcelona. *New Community, 27*(1), 43–58.

Eurostat. (2011). *Migrants in Europe. A statistical portrait of the first and second generation, European Union.*

Farris, S. (2015). Migrants' regular army of labour: Gender dimensions of the impact of the global economic crisis on migrant labor in Western Europe. *The Sociological Review, 63*, 121–143.

Fasani, F., & Mazza, J. (2020). *Immigrant key workers: Their contribution to Europe's COVID-19 response.* IZA Policy Paper No. 155.

Favell, A. (2008). The new face of east-west migration in Europe. *Journal of Ethnic and Migration Studies, 34*(5), 701–716.

Fox, J., et al. (2012). The racialization of the new European migration to the UK. *Sociology, 46*(4), 680–695.

França, T., & Padilla, B. (2017). Reflecting on international academic mobility through feminist lenses: Moving beyond the obvious. *Comparative Cultural Studies – European and Latin American Perspectives, 2*(3), 43–54. https://doi.org/10.13128/ccselap-20825

Fraser, N. (2016). Contradictions of capital and care. *New Left Review, 100*, 99–117.

Friese, M. (1995). Eastern European women as domestics in Western Europe – A new social inequality and division of labour among women. *Journal of Area Studies, 6*, 194–202.

Gallo, E., & Scrinzi, F. (2016). *Migration, masculinities and reproductive labour.* Palgrave Macmillan.

Gorz, A. (1988). *Métamorphoses du travail.* Galilée.

Goss, J., & Lindquist, B. (1995). Conceptualizing international labor migration: A structuration perspective. *International Migration Review, 29*(2), 317–351.

Grigoleit-Richter, G. (2017). Highly skilled and highly mobile? Examining gendered and ethnicised labour market conditions for migrant women in STEM-professions in Germany. *Journal of Ethnic and Migration Studies, 43*(16), 2738–2755.

Haubner, T. (2020). The exploitation of caring communities: The elder care crisis in Germany. *Global Labour Journal, 11*(2). https://mulpress.mcmaster.ca/globallabour/article/view/4090

Hochschild, A. R. (2000) *On the edge: Living with global capitalism* (pp. 130–146). Edited by W. Hutton. Jonathan Cape.

Howard, N. (2019). Neither predator nor prey: What trafficking discourses miss about masculinities, mobility and work. *Anthropology Today.* https://doi-org.ezproxy.mdx.ac.uk/10.1111/1467-8322.12541

Howard, E., & Kofman, E. (2020). *Job quality and industrial relations in the personal and household services sector.* United Kingdom Country Report. https://aias-hsi.uva.nl/en/projects-a-z/phs-quality/country-reports/country-reports.html

International Labour Organisation. (2018). *Migrant workers*. ILO.

IOM/OECD. (2016). *Harnessing knowledge on the migration of highly-skilled women*. IOM/OECD.

Kelly, P. (2012). Migration, transnationalism and the spaces of class identity. *Philippine Studies: Historical and Ethnographic Viewpoints, 60*(4), 153–186.

Kempadoo, K. (Ed.). (2005). *Trafficking and prostitution reconsidered: New perspectives on migration, sex work, and human rights*. Paradigm Publishers.

Kilkey, M., Perrons, D., & Plomien, A. (2013). *Gender, migration and domestic work. Masculinities, male labour and fathering in the UK and the USA*. Palgrave Macmillan.

Kingma, M. (2006). *Nurses on the move. Migration and the global health care economy*. Cornell University Press.

Kofman, E. (2013). Gendered labour migrations in Europe and emblematic migratory figures. *Journal of Ethnic and Migration Studies, 39*(4), 579–600.

Kofman, E. (2014). Towards a gendered evaluation of (highly) skilled immigration policies in Europe. *International Migration, 52*(3), 116–128.

Kofman, E., & Raghuram, P. (2005). Gender and skilled migrants: Into and beyond the work place. *Geoforum, 36*(2), 149–154.

Kofman, E., & Raghuram, P. (2015). *Gendered migrations and global social reproduction*. Palgrave Macmillan.

Kofman, E., Phizacklea, A., Raghuram, P., & Sales, R. (2000). *Gender and international migration in Europe*. Routledge.

Kõu, A., Mulder, C. H., & Bailey, A. (2017). 'For the sake of the family and future': The linked lives of highly skilled Indian migrants. *Journal of Ethnic and Migration Studies, 43*(16), 2788–2805.

Kõu, A. L., van Wissen, L. J. G., van Dijk, J., & Bailey, A. (2015). A life course approach to high-skilled migration: Lived experiences of Indians in the Netherlands. *Journal of Ethnic and Migration Studies, 41*(10), 1644–1663.

Laslett, B., & Brenner, J. (1989). Gender and social reproduction: Historical perspectives. *Annual Review of Sociology, 15*, 381–404.

Leschke, J., & Weiss, S. (2020). With a little help from my friends: Social-network job search and overqualification among recent intra-EU migrants moving from East to West. *Work, Employment and Society, 34*(5), 769–788. https://doi-org.ezproxy.mdx.ac.uk/10.1 177/095001702092 6433

Lévy, F., & Lieber, M. (2009). La sexualité comme ressource migratoire: Les Chinoises du Nord à Paris. *Revue Française de Sociologie, 50*(4), 719–746.

Lutz, H. (Ed.). (2008). *Migration and domestic work: A European perspective on a global theme*. Ashgate.

Lutz, H., & Pallenga-Mollenbeck, E. (2012). Care workers, care drain and care chains: Reflection on care, migration, and citizenship. *Social Politics, 19*(1), 15–37.

Mai, N., Macioti, P. G., Bennachie, C., Fehrenbacher, A. E., Giametta, C., Hoefinger, H., & Musto, J. (2021). Migration, sex work and trafficking: The racialized bordering politics of sexual humanitarianism. *Ethnic and Racial Studies, 44*(9), 1607–1628.

Manalansan, M. F. (2006). Queer intersections: Sexuality and gender in migration studies. *International Migration Review, 40*(1), 224–249.

Maury, O. (2020). Between a promise and a salary: Student-migrant-workers' experiences of precarious labour markets. *Work, Employment and Society, 34*(5), 809–825.

McDowell, L. (2005). *Hard labour: The forgotten voices of Latvian women volunteer workers*. UCL Press.

McDowell, L. (2009). *Working bodies: Interactive service employment and workplace identities*. Wiley.

McDowell, L. (2015). The lives of others: Body work, the production of difference, and labor geographies. *Economic Geography, 91*(1), 1–23.

McDowell, L. (2016). *Migrant women's voices: Talking about life and work in the UK since 1945.* Bloomsbury.

McDowell, L., Batnitzky, A., & Dyer, S. (2009). Precarious work and economic migration: Emerging immigrant divisions of labour in Greater London's service sector. *International Journal of Urban and Regional Research, 33,* 3–25.

McDowell, L., Batnitzky, A., & Dyer, S. (2012). Global flows and the local labour markets: Precarious employment and migrant workers in the UK. In S. Dex, J. L. Scott, & A. Plagnol (Eds.), *Gendered lives: Gender inequalities in production and reproduction* (pp. 123–152). Edward Elgar.

McIlwaine, C. (2020). Feminized precarity among onward migrants in Europe: Reflections from Latin Americans in London. *Ethnic and Racial Studies, 43*(14), 2607–2625.

Meares, C. (2010). A fine balance: Women, work and skilled migration. *Women's Studies International Forum, 33*(5), 473–481.

Moré, P. (2019). 'Here we don't only receive orders'. (Dis)empowering care labour in Madrid and Paris. In M. Amrith & N. Sahraoui (Eds.), *Gender, work and migration. Agency in gendered labour settings* (pp. 30–45). Routledge.

Morel, N., & Carbonnier, C. (2015). Taking the low road: The political economy of household services in Europe. In C. Carbonnier & N. Morel (Eds.), *The political economy of household services in Europe* (pp. 1–38). Palgrave Macmillan.

Morokvasic, M. (1984). Birds of passage are also women. *International Migration Review, 18*(4), 886–907.

Morokvasic, M. (2004). 'Settled in mobility': Engendering post-wall migration in Europe. *Feminist Review, 77*(1), 7–25.

Morokvasic, M. (2007). Migration, gender, empowerment. In I. Lenz, C. Ulrich, & B. Fersch (Eds.), *Gender orders unbound? Globalisation, restructuring and reciprocity* (pp. 69–97). Barbara Budrich Publishers.

Morokvasic, M. (2011). L'invisibilité continue. *Les Cahiers du Genre, 51,* 25–47.

Näre, L., & Nordberg, C. (2016). Neoliberal postcolonialism in the media: Constructing Filipino nurse subjects in Finland. *European Journal of Cultural Studies, 19,* 16–32.

Nowicka, M. (2013). Positioning strategies of polish entrepreneurs in Germany: Transnationalizing Bourdieu's notion of capital. *International Sociology, 28*(1), 29–47.

Nowicka, M. (2014). Successful earners and failing others. Transnational orientation as biographical resource in the context of labor migration. *International Migration, 52*(1), 74–86.

OECD. (2017). Foreign-trained doctors and nurses. In *Health at a glance 2017: OECD indicators.* OECD. https://doi.org/10.1787/health_glance-2017-59-en

OECD/EU. (2018). *Settling in 2018, indicators of immigrant integration.* OECD Publishing, Paris/ European Union.

Oliver, C., & O'Reilly, K. (2010). A Bourdieusian analysis of class and migration: Habitus and the individualizing process. *Sociology, 44*(1), 49–66.

Ortensi, L. E., & Barbiano di Belgiojoso, E. (2018). Moving on? Gender, education, and citizenship as key factors among short-term onward migration planners. *Population, Space and Place, 2018*(24), e2135. https://doi.org/10.1002/psp.2135

Oso, L. (2005). La réussite paradoxale des bonnes espagnoles de Paris. Strategies de mobilité sociale et trajectoires biographiques. *Revue Européenne des Migrations Internationales, 2*(1), 107–129.

Oso, L. (2020). Crossed mobilities: The recent 'wave' of Spanish migration to France after the economic crisis. *Ethnic and Racial Studies, 43*(14), 2572–2589. https://doi.org/10.1080/ 01419870.2020.1738520

Palenga-Möllenbeck, E. (2013). Care chains in Eastern and Central Europe: Male and female domestic work at the intersections of gender, class, and ethnicity. *Journal of Immigrant & Refugee Studies, 11,* 364–383.

Palumbo, L., & Sciurba, A. (2018). *The vulnerability to exploitation of women migrant workers in agriculture in the EU: The need for a human rights and gender based approach.* European Parliament. Policy Department for Citizens' Rights and Constitutional Affairs, Women's Rights and Gender Equality.

Panopio, S. (2010). *More than masculinity: Experiences of male migrant nurses in London.* Working Paper Issue 26, September. Gender Institute London School of Economics.

Parrenas, R. (2000). Migrant Filipina domestic workers and the international division of reproductive labour. *Gender and Society, 14*(4), 560–580.

Parreñas, R. (2001). *Servants of globalization: Women, migration and domestic work.* Stanford University Press.

Parreñas, R. (2020). The mobility pathways of migrant domestic workers. *Journal of Ethnic and Migration Studies, 47*(1), 3–24. https://doi.org/10.1080/1369183X.2020.1744837

Parutis, V. (2014). "Economic migrants" or "middling transnationals"? East European migrants' experiences of work in the UK. *International Migration, 52*, 36–55.

Phizacklea, A. (Ed.). (1983). *One way ticket: Migration and female labour.* Routledge.

Phizacklea, A. (2000). Ruptures. Migration and globalization: Looking backwards and looking forward. In A. In Phizacklea & S. Westwood (Eds.), *Transnationalism and the politics of belonging* (pp. 101–119). Routledge.

Plambech, S. (2017). Sex, deportation and rescue: Economies of migration among Nigerian sex workers. *Feminist Economics, 23*(3), 134–159.

Raghuram, P. (2004). The difference that skills make. Gender, family migration strategies and regulated labour markets. *Journal of Ethnic and Migration Studies, 30*(2), 303–321.

Raghuram, P. (2008). Migrant women in male-dominated sectors of the labour market: A research agenda. *Population, Space and Place, 14*(1), 43–57.

Raghuram, P., & Sondhi, G. (2021). Gender and international students. In C. Mora & N. Piper (Eds.), *The Palgrave handbook of gender and migration* (pp. 221–236). Palgrave Macmillan.

Ramirez, H. (2011). Masculinity in the workplace: The case of Mexican immigrant gardeners. *Men and Masculinities, 14*(1), 97–116.

Rasnaca, Z. (2020). *Essential but unprotected: Highly mobile workers in the EU during the COVID-19 pandemic.* ETUI policy brief 9/2020.

Razavi, S. (2007). *The political and social economy of care in a development context: Conceptual issues, research questions and policy options.* Gender and development paper no. 3. UNRISD.

Rigo, E. (2017). Re-gendering the border: Chronicles of women's resistance and unexpected alliances from the Mediterranean border. *ACME: An International Journal for Critical Geographies, 18*(1), 173–186.

Rostgaard, T., Chiatti, C., & Lamura, G. (2011). Care-migration: The south-north divide of long-term care. In B. Pfau-Effinger & T. Rostgaard (Eds.), *Care between work and welfare in Europe* (pp. 129–154). Palgrave Macmillan.

Ruyssen, I., & Salamone, S. (2018). Female migration: A way out of discrimination? *Journal of Development Economics, 130*, 224–241.

Ryan, L., Erel, U., & D'Angelo, A. (2015). Introduction: Understanding 'migrant capital'. In L. Ryan, U. Erel, & A. D'Angelo (Eds.), *Migrant capital. Networks, identities and strategies* (pp. 3–17). Palgrave Macmillan.

Ryan, L., Lopez Rodriguez, M., & Trevena, P. (2016). Opportunities and challenges of unplanned follow-up interviews: Experiences with polish migrants in London. *Forum Qualitative Sozialforschung/Forum: Qualitative Social Research, 17*(2), Art. 26. http://nbn-resolving.de/urn:nbn:de:0114-fqs1602266

Samaluk, B. (2016). Migrant workers' engagement with labour market intermediaries in Europe: Symbolic power guiding transnational exchange work. *Employment and Society, 30*(3), 455–471.

Sassen, S. (1992). *The global city New York* (1st ed.). Princeton University Press.

Schierup, C.-U., Alund, A., et al. (2014). Migration, precarization and the democratic deficit in global governance. *International Migration, 53*(3).

Schaer, M., Dahinden, J., & Toader, A. (2017). Transnational mobility among early-career academics: Gendered aspects of negotiations and arrangements within heterosexual couples. *Journal of Ethnic and Migration Studies, 43*(8), 1292–1307.

Sert, D. (2016). From skill translation to devaluation: The de-qualification of migrants in Turkey. *New Perspectives on Turkey, 54*, 97–117.

Shire, K. (2015). Family supports and insecure work: The politics of household service employment in conservative welfare regimes. *Social Politics: International Studies in Gender, State and Society, 22*(2), 193–219.

Simpson, R., & Simpson, A. (2018). Embodying dirty work: A review of the literature. *Sociology Compass, 12*, 1–9.

Sondhi, G., & King, R. (2017). Gendering international student migration: An Indian case study. *Journal of Ethnic and Migration Studies, 43*(8), 1308–1324.

Sondhi, G., Ragurham, P., & Herman, C. (2018). Skilled migration and IT sector: A gendered analysis. In *India migration report 2018 – migrants in Europe*. Routledge.

Sowa-Kofta, A. (2017). Central and Eastern European countries in the migrant care chain expert group meeting on *"care and older persons: Links to decent work, migration and gender"* December 5–7. https://www.un.org/development/desa/ageing/wp-content/uploads/sites/24/2017/11/Sowa-Kofta_PP_EGM_Migrant-Care-Chain.pdf.

Standing, G. (2001). Care work: Overcoming insecurity and neglect. In M. Daly (Ed.), *Care work: The quest for security* (pp. 15–32). International Labour Office.

Standing, G. (2011). *Precariat*. The new dangerous class.

Stournara, N. (2020). *'Paradigmatic workers': Sociologies of gender, class and ethnicity in the labour experiences of Albanian and ethnic Greek Albanian women cleaners at two Greek public hospitals*, PhD thesis Middlesex University London.

Thompson, M., & Walton-Roberts, M. (2019). International nurse migration from India and the Philippines: The challenge of meeting the sustainable development goals in training, orderly migration and healthcare worker retention. *Journal of Ethnic and Migration Studies, 45*(14), 2583–2599.

Triandafyllidou, A., & Isaakyan, I. (Eds.). (2016). *High skill migration and recession: Gendered perspectives*. Palgrave Macmillan.

Tyldum, G. (2015). Motherhood, agency and sacrifice in narratives on female migration for care work. *Sociology, 49*, 56–71.

Van Hooren, F. (2012). Varieties of migrant care work: Comparing patterns of migrant labour in social care. *Journal of European Social Policy, 22*(2), 133–147.

van Bochove, M., & Engbersen, G. (2015). Beyond cosmopolitanism and expat bubbles: Challenging dominant representations of knowledge workers and trailing spouses. *Population Space and Place, 21*(4), 295–309.

Vattinen, T. (2014). Reading global care chains as migrant trajectories: A theoretical framework for the understanding of structural change. *Women's Studies International Forum, 47*(Part B), 191–202.

Villosio, C. (2015). *Migrant workers in the healthcare sector in 5 European countries:: A quantitative overview from EU-LFS*. FIERI and LABOR.

Von Bose, K. (2019). Cleanliness, affect and social order: On agency and its ambivalences in the context of cleaning work. In M. Amrith & N. Sahraoui (Eds.), *Gender, work and migration* (pp. 46–61). Routledge.

Vosko, L. F. (2010). *Managing the margins: Gender, citizenship, and the international regulation of precarious employment*. Oxford University Press.

Walton-Roberts, M. (2021a). Intermediaries and transnational regimes of skill: Nursing skills and competencies in the context of international migration. *Journal of Ethnic and Migration Studies, 47*(10), 2323–2340.

Walton-Roberts, M. (2021b). Bus stops, triple wins and two steps: Nurse migration in and out of Asia. *Global Networks, 21*(1), 84–107. https://doi.org/10.1111/glob.12296

Weinar, A., & Klekowski von Koppenfels, A. (2020). *Highly-skilled migration: Between settlement and mobility*. Springer.

Williams, F. (2018). Care: Intersections of scales, inequalities and crises. *Current Sociology, 66*(4), 547–561.

Wills, J., May, J., Datta, K., Evans, Y., Herbert, J., & McIlwaine, C. (2010). *Global cities at work*. Pluto.

Wolkowitz, C. (2006). *Bodies at work*. Sage.

Wojczewski, S., Pentz, S., Blacklock, C., Hoffmann, K., Peersman, W., Nkomazana, O., & Kutalek, R. (2015). African female physicians and nurses in the global care chain: Qualitative explorations from five destination countries. *PLoS One, 10*(6), e0129464. http://www.ncbi.nlm.nih.gov/pmc/articles/PMC4466329/

Yeates, N. (2009). *Globalizing care economies and migrant workers*. Palgrave Macmillan.

Yeates, N. (2010). The globalization of nurse migration: Policy issues and responses. *International Labour Review, 149*(4), 423–440. https://doi.org/10.1111/j.1564-913x.2010.00096.x

Yeoh, B., & Willis, K. (2005). Singaporeans in China: Transnational women elites and the negotiation of gendered identities. *Geoforum, 36*(2), 211–222.

Chapter 4
Transnational Families, Intimate Relations, Generations

Chapter 3 examined the gendered nature of a migrant division of labour. In this chapter we turn to family migration, traditionally associated with women as dependents and followers of men. The term is used to categorise the international movement of people who migrate due to new or established family ties. People moving for family reasons constitute the largest group of migrants entering OECD countries, ahead of labour and humanitarian migration (OECD, 2019). To move for family reasons may encompass an array of different kinds of migration trajectories, from the adoption of a foreign child to family members accompanying migrant workers or refugees, as well as people forming new family units with host country residents or family reunification (when family members reunite with those who migrated previously).

While international migration had traditionally been equated with the movement of men, women were depicted as the followers in what was seen as secondary migration. However, the growth of female labour migrants in domestic work, care work, and nursing, meant that women too became the sponsor of husbands, children, and parents. In the early 2000s, Bryceson and Vuorela (2002) drew attention to transnational families in Europe and countries of origin, leading to the consolidation of family migration as a field, which has burgeoned in the ensuing years.

In this chapter we first outline the growing interest in family migrations, theoretically and in policy terms, which has generally sought to restrict flows and been based on particular gendered representations of family members and women in particular. Recently marriage migration has generated interest, though the area of greatest expansion has been that of transnational families.

4.1 Developing Family Migration

Starting in the late 1980s, theoretical and methodological research on family migration emerged as a subject of scholarly work (Boyd, 1989). The role of the family in internal migration and a number of country case studies of family reunification to the

A. Christou, E. Kofman, *Gender and Migration*, IMISCOE Research Series, https://doi.org/10.1007/978-3-030-91971-9_4

US were published (see *International Migration Review*, 1977, 1986). In Asia, too, family migrations, and especially marriages, gave rise to articles in the *Asia and Pacific Migration Journal* (1995, 1999). European research, however, lagged behind as family migration drew less attention than labour migration due to its association with dependency upon a primary migrant, and its majority of women and children. Even so, as labour migration grew in numbers, so too did family migration increase as a result of increasing family reunifications, a trend that also characterises southern European countries since the 1990s (Ambrosini et al., 2014). Hence, over the past two decades, we see consistent growth of publications about family migrations in relation to its different forms, the experiences of different family members, familial strategies, and the formation of transnational families.

While earlier studies focused on countries of destination and often assumed that migrants wished to bring their family members with them, more recent studies take a more nuanced and critical view of migratory processes and question the desire to complete family reunification. Theoretically, studies have adopted a *migration systems approach* in which all forms of migration (permanent, temporary, circular, return) occur simultaneously. Increasingly, studies are multi-sited and transnational, in which people, services, and cultural and social practices circulate between places, underscoring the interdependency between the mobile and immobile to ensure successful migration outcomes (Bélanger & Silvey, 2019).

Although family-related reasons play a significant role in intra-European migration (depending on the data source), this perspective has been somewhat neglected among researchers. Some of the difficulties of identifying family-related movements arise from the fact that individuals are often not counted as such because they do not hold a residence permit under this category and because restrictions on movement for family reasons do not apply to the same extent for EU nationals. Yet, large-scale migration from Eastern Europe post-EU enlargement in 2004 drew attention to the family strategies deployed by Polish migrants in their migration to and settlement in western Europe and relationships with their homeland (Ryan et al., 2009).

However, it has not only been academic interest in family migration that contributed to the growth of publications and comparative European projects. The Europeanisation of migration policy from Tampere onwards gave rise to the adoption of the Family Reunification Directive 2003/86 EC in 2003 (adopted by all Member States except Denmark, Ireland and the UK). The Directive outlines the minimum rights third-country nationals should have in reuniting with a family member living in an EU member state, but does not address the situations of third-country nationals who are family members of an EU citizen.

The Directive also provides more favourable rules for refugees. It has been progressively adopted over a number of years by old EU states as well as the new enlargement states. The Commission has monitored (2008, 2014, 2019) the implementation of the Directive while the European Migration Network (EMN, 2017) has produced reports on issues and problems regarding family reunification and related issues. In part, concerns about family migration are due to the fact that, for the past 30 years family reunification has been one of the primary drivers of immigration to the EU. In 2017, 472,994 migrants were admitted to the EU-25 on grounds of family reunification, or approximately 28% of all first permits issued to third-country nationals in the EU-25. And although, employment has since 2016 become the main reason for permits due in part to large-scale emigration from the Ukraine, it still accounts for a substantial percentage – in 2019 (Eurostat, 2020).

While earlier research focused on family reunification of migrants and co-ethnic marriages, more recent research has turned to how family migration policies define the acceptable family and permissible intimate relationships which includes a range of family members and familial and kin relationships, but also other affective relationships (e.g., love and marriage, parenting of children, and parental care) (Mai & King, 2009). It should be noted that 'family' for the purposes of migration policy was conceptualised as a traditional nuclear family comprised of a married couple and dependent children under 18 years of age.

Migrants benefitting from family migration regulations are expected to demonstrate they have the capacity to be productive, comply with acceptable cultural practices, and not be a burden on the welfare state. The complexity of how family members contribute to the social reproduction of the family tends to be given little attention in the migration literature. Rather, attention is paid to the nuclear family in immigration legislation while the roles of parents and other kin are marginalised (see transnational families below).

The right to family reunification and formation – income, other resources such as housing, integration conditions – has generated inequalities. Family reunification policies are most restrictive in northern countries, as they align the conditions for sponsorship increasingly with economic conditions for labour migration, especially the high-income requirements in a number of countries (Kofman, 2018; Pellander, 2021; Staver, 2015), which has rendered class (see also Chap. 2) more significant in the stratification of access to family life together with gender inequalities.

Marriage migration (D'Aoust 2013) as a separate issue to family reunification is a more recent area of study within family migrations. In Europe, research tended to be focused, initially, on marriages between co-ethnics such as Turks, Moroccans or Pakistanis marrying with someone from their homeland and seen as a problem for integration of the migrant in the receiving society (see Chap. 6). Cross-border marriages between a wider range of nationalities than co-ethnic as a means of migrating legally and acquiring citizenship have begun to receive more attention. Such migrations raise questions about regulation of who belongs and who deserves citizenship (Fresnoza-Flot & Ricordeau, 2017) (see Chap. 6) and has become increasingly politicised (Moret et al., 2021).

Intra-European binational couples have been surprisingly under-studied (De Valk & Medrano, 2014) due in part to the assumption that intra-European mobility is primarily driven by work reasons. However, Migali and Natale (2017) found that familial reasons are nearly as significant as work motivations. For many individuals the movement for familial and intimate reasons represented a second mobility, following an initial move for education or work (Gaspar, 2012). Having the privilege of EU citizenship and free movement rights, couples do not have to marry, but may cohabit. However, same sex marriages are only recognised and performed in some northern, western and southern European states (16 states as of July 2020). A few others, such as Croatia and Hungary, also recognise same sex partnerships.

In the next section, we turn to the growing interest in transnational families as the outcome of large-scale migrations, some of which have permitted the reconstitution of the family in one place, whilst others have led to separated families, especially amongst those who do not meet the requirements for crossing borders or fulfilling integration criteria (see Chap. 6).

4.2 Transnational Families: Concepts, Generations, Relations

As a historical phenomenon of social organisation under globalisation and an inevitable result of migration, the emergence of transnational families marks the development of 'familyhood' relations (Bryceson & Vuorela, 2002) across national borders. Transnational familyhood offers the opportunity to negotiate family life through intersecting individual and familial objectives but with structural constraints as crucial as mobile opportunities can possibly become (Fesenmyer, 2014). Transnational families are also organised under the rubric of sustained familial ties, kinship networks and fluid relations spanning two or multiple nation states. There is a clear realisation here that migration and other kinds of mobilities are not unidirectional and do not focus on permanency of settlement and relocation with a country of destination as the core objective. The more crucial realisation is that under such arrangements of long-distance transnational familyhood, the maintenance of familial connections, kinship feelings, collective belonging to a family unit with notions of shared welfare, shared responsibilities, caring arrangements, participation in social reproduction and group consumption are all part and parcel of such relations (McCarthy & Edwards, 2011).

As Bryceson (2019: 2) confirms: 'transnational families are not new but they are definitely more numerous now than ever before. They are an evolving institutional form of human interdependence, which serve material and emotional needs, in the twenty-first century's globalising world. The transnational family constitutes a multi-dimensional spatial and temporal support environment for migrants, as well as providing motivational impetus for migration'. The breadth and depth of family relations will fluctuate according to the specific needs, be that economic or emotional, as will the strategic incentive to extend the social networks within or beyond the ethnocultural or religious parameters that might define the boundaries of the group. It is then obvious that the concept of 'transnational familyhood' is framed around how borders are perceived, understood, experienced, negotiated and socially constructed by migrants.

Individual migrants and their families don't just navigate psychic and emotional borders in their mobile lives or in pursuit of settlement. They frequently contend with a number of logistical, legal, bureaucratic and pragmatic 'border-crossings' such as residency, citizenship, immigration and labour policies etc. often in the midst of sudden geopolitical shocks and wider global turbulences. Their circumstances might give navigational direction but the wider global scope engulfing their acts and actions is what inherently defines most outcomes. As Bryceson (ibid) argues: 'Since the turn of the millennium, a tendency for transition from blurred to brittle borders has gained momentum in the European Union and North America. 'Blurred borders' refer to migrants' low risk border crossings and light regulation of their visits, affording relative openness to migrants seeking legal residence and citizenship in the receiving countries. 'Brittle borders' represent the opposite, involving physically and legally hazardous crossings with enforcement of stringent restrictions on temporary as well as permanent residence and remote possibility or impossibility of migrants gaining citizenship or family reunification'.

Transnational families might seek to secure their immigration status and citizenship in the destination country, or, they might further enhance affective locational nodes with the ancestral homeland or the country of origin. However, there is a third way in the option of maintaining both a new livelihood niche setting in the country of destination while also continuing the exchanges and family dynamic communications back 'home'. A variety of factors (e.g. class, educational capital, networks etc.) might shape the prioritisation of the children's educational and cultural integration in the destination country and as a result more 'hybridised' transnational identities might emerge for the offspring and subsequent generations.

Generations are also integral to understanding dynamics, constraints and opportunities for transnational families. Before we disentangle the concept of 'generation', it is important to clarify that within the varying typologies of the 'migrant' category, transnational families do evolve. That is, be it economic or irregular migrants and refugees or the more privileged status of highly skilled mobile professionals or scientific and student migration, transnational familyhood becomes relatively central among those groups when it comes to motivations with regard to settlement, employment or potentially return. All indicative migrant categories above might compete for welfare and employment opportunities but also face similar dilemmas as regards their potential settlement or return migration. Additionally, some of the drivers shaping their trajectory might be similar, such as experiences of racism, exclusion, discrimination as regards housing, work and other social encounters. In the case of those envisaging returning to their country of origin, often, the maintenance of transnational ties becomes strategically useful in keeping aware of the political, social and economic situation in the sending country. As a result, the propensity, intensity and desire for transnational engagement can only be fully understood in the context of each national group and geographic context (Bloch & Hirsch, 2018; Carling & Pettersen, 2014).

A similar kind of fluidity exists when we try to operationalise the concept of 'generation'. As a migration chronotope, a spatially and temporally situated phenomenology, and, the ontological, imaginary and state policy parameters within which we emplace genaeologies, the concept of 'generation' is multidimensional and complex (King & Christou, 2010; Christou & King, 2015). That is, 'generation' is not simply a matter of linear, temporal and geographic origins that have a neat trajectory from the zero, first, to second and subsequent generations. The evolving identities of those offspring born in the country of destination are even further challenged by the 'tyranny of ethnic consciousness' of each parent, the family context experiences, their schooling and of course their own sense of agency and belonging. As a result, we slip into more complicated percentages when beyond 'first' and 'second' generation migrants we talk about the 1.25; 1.5; 1.75 (Portes & Rumbaut, 2006) or even 2.5 generations (Ramakrishnan, 2004) mostly in the US literature but with increasing interest in the rest of the international literature (Safi, 2018).

One alternative way to move beyond the chronotype limitations of confining groups to 'generations' is to utilise the concept of 'cohort' in conceptualising the succession of migrant waves as 'diasporic' (Berg, 2011) but also sociologically in terms of genealogical spheres of historical contexts and group accounts (Kertzer, 1983). The generational element is also integral to contemporary research on transnational grand-parenting in the digital age (Nedelcu, 2017; Janta & Christou, 2019) in understanding the role of the 'zero generation', in this case being the grandparents, who although might not have migrated do engage in transnational grand-parenting (Nedelcu, 2017). Here, the dynamics of transnationalism become enmeshed with multiple modalities, not only three generational relations (parents, children, grandparents) but also implications of technologically mediated co-presence, leisure travel and hosting in shaping new social practices. Transnational family practices in the digital age are shaped by new media and polymedia regimes which in a sense reinvent social reproduction practices (Madianou & Miller, 2012; Baldassar et al., 2016). This kind of 'cyber-transnationalism' with its digital webs of connectedness defines new intra and inter-generational possibilities of collaboration and conflict. The motivation however remains to maximise contact and in the case of transnational grand-parenting to enhance caring circuits. We will revisit this aspect of generations as regards parenting, ageing and caring in the next two sections that follow.

However, beyond the digital context, affective negotiations remain intact, as even in the case of digital transnational relations, elements of disagreement and tensions render dysfunctionality as a normalising aspect of transnational communication for families. This way 'doing family' (Christou & Michail, 2015) and 'doing caring' transnationally (Baldassar, 2014) is subject to mainstream normalising processes of emotional labour and relational negotiations.

At the same time, while the strategic reference for migration is often perceived as individualistic in its motivation and execution, both motives and outcomes are frequently linked to the family. The scope and dynamics of parenting as well as the impact on childhoods is inextricably linked to the context of such which in the case of transnational migration poses additional challenges. In the next section, we cover some of the core themes emerging in the literature on family studies of transnational parenting.

4.3 Transnational Parenting and Childhood

The proliferation of studies on transnational families from a variety of inter/disci-plinary perspectives over the past two decades (Zontini, 2010; Carling et al., 2012; Menchavez-Francisco, 2018) might have engaged with the wider topic of transna-tional familyhood but less frequently with gendered impacts on the emotional well-being of parents and children (Jordan et al., 2018; Caarls et al., 2018).

It has been observed that 'a relatively neglected trigger of migrants' transnational involvement is their position in the life course and, at a macro level, the stage of a migration cycle they belong to' (Boccagni, 2012: 42). Parenting and grand-parenting can be considered as roles pivotal to both the life course and the migration cycle that migrants experience. Empirical studies focusing on transnational (grand) parenting practices help us understand transnational families as a 'living strategy' (Kofman, 2019) relevant to a broader spectrum of transnational sociologies of the family. Moreover, the focus on how family migration can be disruptive to gender roles (ibid) illuminates new directions for the study of gender and migration.

Family relations transnationally are both complex and fluid especially when it comes to young people and those in a vulnerable context such as unaccompanied asylum-seeking youth (Devenney, 2020). However, an additional layer of analysis observed in this body of works is that of perceiving unaccompanied children and young migrants as victims and by extension using language mirroring saviour mentality and saving mechanisms. These migrants are perceived primarily in terms of their physical dislocation from family members (in legal terms the status of 'unaccompanied' is framed that way) and by extension have focused on their sense of loneliness, trauma, separation, vulnerability, victimhood, fragility, passivity and neediness (Herz & Lalander, 2017; Devenney, 2020). As a result, the wider discourse of 'saviour social institutions' readily aware of their role in stepping in to rescue youth who are seen as totally severed from any kind of emotional and logistical transnational networks tends to proliferate, while the actual transnational connections and family relations of those young people are mostly overlooked in otherwise fascinating studies.

Moving away from a solely 'Western-oriented' perspective, for instance, in a study by Kõu et al. (2017) on the Indian migrant context to the Netherlands and the UK, the presence of parents and extended family in the constellation of migration trajectories highlights that life course events of linked lives shaped by key elements of care-giving and care-receiving becomes a setting for negotiating social norms and expectations. This particular study explored how family members facilitate through social norms migration and how such norm negotiations involve expectations for the provision of care-giving. In other contexts, such as different Southeast Asian households, the emergence of transnationally stretched families and global householding practices at different parameters of the care chain could give rise to different kinds of gender politics influencing the social provisioning of everyday and generational care (Yeoh, 2016).

This brings us to some of the useful conceptual tools or heuristic explanatory frameworks that help us explain experiential transnationality in migrant familyhood. For example, Goulbourne et al. (2010) have emphasised the usefulness of different spheres within which transnational family lives unfold, be that in socio-economic or politico-cultural parameters which might overlap or even intersect. They contend that such an approach contributes another critical lens in disrupting and de-centring the 'normative' status of the 'the family' (ibid).

Another conceptual approach that has been advanced by Anthias (2008, 2009, 2012) as providing a more integrated analysis in terms of locations and identities in moving away from specific 'groups' or 'categories' (gender/ethnicity/class) is that of an 'intersectional translocational positionality' (for an elaboration refer to Chap. 2). This approach argues for a more intersectional framing that pays particular attention to social locations and processes than social divisions. While the approach appears compelling, it might also signal an indirect tension between understanding the particularities of specific figurations shaped by social categories when one of the intersectional formations appears more pronounced in specific social circumstances (be that age, gender, class, ethnicity, race, generation, ability) driving those experiences in time and space.

A third conceptual approach to understanding transnational childhood and parenting is that of incorporating an affective lens. Despite the invisibility of migrant children as active members with agency in transnational families, a focus on 'emotional labour' can address them as actively engaging with transnational socialisation processes (King et al., 2011; Zeitlyn, 2012). Indeed, Mand (2015) suggests that the focus on the emotional labour of children as social actors can illuminate the nature of their agency, their emplacement and power relations performed as part of the transnational familial habitus. The lens of emotions also highlights that within transnational mobility an 'affective habitus' (Christou & Janta, 2019) underlines the significance of emotional encounters with transnational kin networks in refining social practices. It is intriguing to keep in mind that the emotional labour performed by migrant children, embodied and active, is a clearly agential effort by children to capture and acquire the appropriate socio-cultural capital to legitimise their belonging (King et al., 2011; Zeitlyn, 2012). While there is ambivalence in emotions and transnational lives, acts of agency by children are often overlooked as their experiences are not always analysed as socialising processes enmeshed with mobility. Hence, emotional experiences and embodied performances of transnational childhood should be viewed as constellations of socialising activity in mobility.

The agency of children's practices in transnational familyhood transforms but also reproduces transnational social fields and as a result children actively develop and negotiate relationships with family members across transnational spaces (Gardner, 2012). Children exhibit active agency into transnational migration decisions (Ní Laoire et al., 2010; Cebotari et al., 2017) but it is relevant to examine changes to this as regards cultural factors, age, personal development. One core aspect of this discussion is to unpack the relationship between parental migration and child well-being but also the various dimensions in transnational parenting, some of which are discussed in the box below drawing on key empirical global evidence.

Box: Transnational Parenting: Some Indicative Research Results

Insights on reframing transnational mothering when it is seen as being 'troubling' when induced by the act of migration since children are left behind has been explored by Irena Juozeliūnienė and Budginaitė (2018) who combine interviews with transnational mothers over a period of a decade, but, also includes analyses from 79 articles on transnational families published over a 9 year period, along with national press and media sources in Lithuania. The analysis of the transnational migrant mother portrayals reveals that mothering is scripted and as an indication of agency, migrant mothers 'normalise' troubling narratives of their mothering and discredit those, thus bringing new *meanings to migrant mothering* performativities. These new storied accounts of mothering demonstrate agency in shifting previous ones and crafting alternative mothering performances. Hence, migrant mothering while being contested is also constructed and reinvented.

Research by Kufakurinani et al. (2014) regarding 'transnational parenting and the emergence of "diaspora orphans" in Zimbabwe' draws from interviews with adults providing childcare for left behind youth and children in Zimbabwe, including single parents, grandparents, siblings and care professionals. The research adds innovative debate strands from an *intersectional* aspect in focusing on the middle class, gender and the life stages. It also fosters more refreshing debates on diverging the analysis from poverty entrapment and subsequent labour migration during the crisis period in Zimbabwe, to issues of morality and emotionality on changing parenting practices and social discourses. One of those is on the emergence of 'diaspora orphans' as a negative stereotyping of moral parameters and a crisis of wealth when middle class families migrate but leave the children behind. It addresses aspects of *transnational parenting* from the angle of the carers and issues of intimacy in relationships and childcare.

Another politically charged aspect of transnational parenting has to do with changing gender ideologies when these are a result of migrant mothers engaged in labour migration (see Chap. 3) and have to redistribute caring responsibilities and carework required for children. The research by Lam and Yeoh (2018) draws on both quantitative and qualitative interviews with returning migrants and left-behind kin from communities in the Philippines and Indonesia. There are a number of emerging implications for left-behind fathers when mothers are absent.

Similarly to 'doing gender', 'doing care labour' not only reflects existing social hierarchies, it also reproduces or even exacerbates them. The evolving of new father-driven caring in the absence of migrant mothers in the Southeast context of Lam and Yeoh's (2018) research creates 'new package deals' of *reconstructing masculinities*.

These point to harnessing elements of responsibility, capability, adaptability and control in retaining on the one hand a sense of power and pride but also adjusting to the circumstances for successful provision of care (ibid: 113–114). Masculinities of fathering practices are thus strongly anchored on enduring obstacles while safeguarding self-image. These conclusions mirror to a great extent those findings of Pasura and Christou (2018) who have moved away from conceptualisations of black transnational forms of masculinities in perpetual crisis and have explored diverse notions of such as being challenged, reaffirmed and reconfigured. Moreover, beyond major global structural changes, both these research studies point to the sociologies of family relations as a livelihood shifting arrangement that combines the messiness of intimacy, affectivity, bonds, normative framings and pragmatic decision-making. In a sense gender identities and family roles appear to be blurred or undone when fathers become both mothers and fathers and mothers increasingly breadwinners in the household division of labour.

To an extent, these reconstructed and negotiated figurations of gendered scripts point to the mouldable capacity of roles as shaped by economics and not just socio-cultural framings. That is, as Lam and Yeoh (2018: 114) suggest: 'In the era of migration and family survival, "doing family" may thus become more important than "doing gender"'. While the scale, intensity and practices in transnational parenting will differ across generations and geographies shaped by social locations, by and large, they illustrate that *gender* is an important parameter in the process. There are instances where its meaning is re-discovered in being re-invented and reconstituted, but above all, it is a *defining dynamic* for families, relations and identities.

Interesting findings emerge from a study by Gassmann et al. (2018) using household survey data from Moldova and Georgia to measure and compare multi-dimensional aspects of child well-being while 'left behind'. Gassmann et al. (ibid) in their study are challenging normative accounts which claim that parental migration can have negative, harmful or destructive impacts on children's well-being outcomes, and, instead focus on dimensions of wellness as regards education, health, housing, safety, communication and emotional well-being. Their study operationalises well-being in a holistic and multidimensional framework that refutes opposing narratives of transnational parenting as dysfunctional.

Gender ideologies also intersect with layers of dynamism in how caring unfolds given the transnational habitus that children experience. In the instance of the global South, Kofman and Raghuram (2012) alert us to a dynamism and diversity of caring arrangements and point to the importance of empirical studies taking on board a wider interplay of the household that incorporates elements of communities, the state, financial markets and intra-family relationships. In the case of the research with Zimbabwean children, it appears that money for instance, undermined the impact of 'authority in families' (Kufakurinani et al., 2014: 127) and actually shaped the quality of caregiving which was dependent on remittances received when there were no complaints of their service provided. As an extension of this however, children appear to have a degree of agency in not accepting or resisting the authority of caregivers since the latter for fear of losing their source of income would not contest the children's behaviour. Wider external and outsider opinion would highly criticise the above strategic lax child control on the basis of personal interest rather than appropriate child developmental practices. Bluntly put, the employed caregivers seemed to 'care' more about the money they received rather than properly caring for the left behind children.

As Christou and Michail (2015: 72) observe in their research on migrant mothers, there is a 'continuous complexity that the dichotomy of private and public invokes' as they seek to understand how those spaces intertwine and intersect with mothering practices. Their research with women in their 40s and 50s from Albania, Bulgaria, Romania and Poland who have migrated to Greece and within a transnational habitus are raising children there is premised on viewing migrant mothers as co-producers of 'inclusive socialization of the second generation as agents of intercultural citizenship' (ibid).

Indeed, as Kufakurinani et al. (ibid: 135) suggest, research on transnational families with a 'focus on the nodes of the "care triangle" that are often overlooked in studies of transnational parenting can be particularly revealing – i.e. the views of the foster parental caregivers and those in positions of authority over such children within schools and communities, as well as the children themselves'. Hence, it is integral to also view children as agents of change, often involved in the migrant transnational caring context and not simply as passive recipients of migration, parenting decisions and practices.

Cross-border family arrangements linked to the increase of 'mother-away' families can also lead to questions of 'moral accountability' of migrant mothers regarding the actual well-being of their children (Juozeliūnienė & Budginaitė, 2018) as linked with the significant increase of international migration patterns of Lithuanians following the country's 2004 accession to the EU. Juozeliūnienė and Budginaitė provide some core insights to these issues based on interviews with transnational mothers over a decade (2004–2014) combined with media depictions of transnational families. Such long-distance mothering practices and cross-border family arrangements are considered as problematic in being designated as 'troubles' for the children who remain in Lithuania while their mothers migrate to work overseas. There are gendered ways here through which applications of 'moral standing' as regards childcare are enacted in classifying mothering practices through narratives of parental ethical stances. These are situated within discussions of how transnational parenting can become politicised and policed scripts.

At the same time, there is research (Haagsman, 2018) that suggests links between transnational family life and negative outcomes for job prospects when comparing Angolan parents with 'left behind' children and those who live in the Netherlands with all their children. The findings, cross-cutting migration studies and organisational psychology through mediation analysis, indicate that transnational family life has a direct impact in increasing migrant parents changing jobs due to low levels of happiness. This is an integral aspect of a more holistic study of transnational family relations in redirecting attention of how transnational family life circumstances affect the employment potential outcomes for parents, especially as labour prospects and financial returns are seen as a core objective to the choice of living transnational lives. So, where transnational migration and family studies have been preoccupied with the affective impact for children, similar investigations regarding parental transnational working lives have been scarce. In the next section, we'll explore additional affective realms of transnational lives, namely those surrounding intimacies and sexualities.

4.4   Transnational Intimacies and Sexualities

The expansion toward more diverse forms of intimacy in familial relations as well as issues of sexuality are imperative for both research and policy in returning to the moral sense of facilitating less economic driven regulation and more socially just reunification, at present, when it comes to family migration (Kofman, 2019).

Transnational intimacies and sexualities are important as gendered practices and inextricably linked to global mobilities. However, the study of transnational intimacy, sexual relations and sexual migrant identities, has, for a long time, been confined to heteronormative parameters defining relationships and families under such a unidirectional gaze. Although a decade or so ago Mai and King (2009) argued for a combined 'sexual' and 'emotional' turn in mobility studies in underscoring the intersectionalities of these two dimensions and their grounding in more productive queer theory driven research (see Chap. 2), there has been slow progress in global investigations informed by those intersections. Queer mobilities account for the emotional and embodied dimensions as shaped by desires and intimate attachments (Gorman-Murray, 2009) and the possibilities of transnational feminist queer research can contest configurations of power and hierarchies of the Global North/ Global South (Browne et al., 2017). Engagements with the multiplicity of the politics of place, as well as across geographical locations, highlight not only spatial nuances but also brings researchers into dialogue with diverse flows and boundaries.

Yet, as highlighted in the queer migration literature, homemaking in diverse home spaces tends to be negotiated in spaces of liminality, dislocation and opposition especially to homonationalism and heteronormativity (Luibhéid, 2008; Mole, 2016; Wimark, 2019). At the same time, some of those tensions have been used as a springboard for research focusing on the potential benefit of 'queer diaspora' as a heuristic device to think about identity, belonging and solidarity among sexual minorities in the context of dispersal and transnational networks (Mole, 2018). Moreover, there is a continued call to apply gender analysis when studying post-migration experiences of lesbians and gay male immigrants (Fournier et al., 2018).

As a research instrument, a transnational sexualities approach is committed to being theoretically, epistemologically and ethically self-reflexive to the co-production of knowledge and the de-centring of established categories. It is an inclusive approach in embracing largely marginal populations in the process of knowledge production, in intersectional terms, (for instance, working-class queer disabled religious ethnic minority sexualities) queering and questioning social categories, politics, practices and ideologies that reproduce exclusion. Hence, transnational sexualities approaches provide a critical lens to explore connections and circulations of sexual subjectivities, sexual discourses and mobile practices across two or more national contexts. Within the conceptual parameters of transnationality, sexualities and intimacies are understood as sexual formations imbued with mobility meanings in a terrain whereupon local, global and national hegemonies and politics intersect.

Research on transnational sexualities and intimacies maintains a sustained attention to both cultural and historical figurations within circuits of transnational mobility, relational dynamics in space, time and embodied performances. By extension, transnational sexualities and intimacy studies need to locate their analyses within contexts of cultural, political and historical connections while being informed by broader questions highlighting issues of globalisation, modernity, development, capital, nationalisms, colonialisms, imperialisms, racialisations, etc. as departure points to problematise discourses, otherwise reified, naturalised or normalised.

Contestable and 'notoriously obscure, due to its conceptual complexity, historicity and political situatedness' (Chattopadhyay, 2018: 149), the concept of borders is also compelling and pivotal to challenging Eurocentric representations of migrant intimacies in understanding everyday lives of migrant trajectories from the Global South to Europe. Such insights forge parallels of the biopolitics of borders with the geopolitical histories of colonialism and imperialism in Europe. Again, these (often critical and feminist informed) conceptualisations deepen understandings of migrant experiences, identities and intimacies. In various social spheres there is an increased emergence of experiences of everyday borderings which are differentiated according to situated circumstances and social positionings of migrants (Yuval-Davis et al., 2018).

Indeed, borders and bordering lives have become 'a flourishing research agenda on everyday geographies and ontologies of personal (in)security' (Vaughan-Williams & Pisani, 2018: 1). While the recent 2015 'Mediterranean Migration Crisis' has brought to the fore and made visible a number of long-standing border regime crises (ibid), there is also a body of works that has examined the border as a social construction within historical, geopolitical and socio-cultural underpinnings always undergoing varying processes of transformation (Johnson et al., 2011) (see Chap. 5). It is argued that marriage migration 'is integral to many transnational communities, that is, those that have established themselves in several countries across the world. In other cases, marriage migration may represent a first migration step which creates a potential for future links between countries' (ibid, p.1).

Examining the intersections between marriage, sexuality, migration and citizenship contributes to understanding all concepts outside their singularity as a social phenomenon or social status (Chauvin et al., 2021) and underscores that these need to be located within the larger political economy and historical trajectories encountered in particular regions (Ibrahim, 2018). For instance, in 'the context of South Asia, contemporary cross-border migration builds on networks of trade, labour, and marriage that have endured across partition, and is a potent reminder that new borders drawn up in 1947 or 1971 were not the definitive hiving-off of territory so much as the inauguration of new regimes of citizenship and border management on the part of the state, and of new expressions of identity and belonging for its citizens' (ibid: 1665). It becomes clear that cross-border marriage mobility should be viewed within sociological and historical parameters, discursive conditions and negotiations of belonging and citizenship. Hence, transnational 'marriage-scapes' gender our understanding of contemporary transnational migration by highlighting the effects of border-crossings on familial and gender roles.

A redirection to the study of 'intimate mobilities' has occurred because 'much migration research remains desexualized and overlooks emotional and intimate relationships' (Groes & Fernandez, 2018: 1). Reconfigurations of gender relations are also complex in transnational marriage contexts and those can revert to traditional roles, be amplified or even reversed and undermined depending on the parameters of socio-cultural integration (Charsley, 2012; Charsley et al., 2016).

But integration and gender are also pertinent in the discussion of migrant care and ageing in transnational fields (Zontini, 2015). Findings by Zontini (ibid) from her long-term study of transnational ageing Italian migrants in the UK, reveal that community and belonging enhance successful ageing but above all, strength and reciprocity of bonds with co-ethnics locally and transnationally showed a sense of well-being linked to those experiences. Nevertheless, overall, findings of studies on transnational caring of ageing migrants demonstrates that there are many challenges involved, from negotiating the expectations and obligations of caring to issues of loneliness and trauma when those expectations and obligations are disrupted by migration of the children or even the complications of 'crises' phases in the care responsibilities of transnational families (King & Vullnetari, 2006; Baldassar, 2007, 2014).

Useful terminology to conceptualise the complexities, fluctuations and networkings regarding ageing care are, for instance, the one proposed by Baldassar and Merla as 'circulation of care' referring to the 'reciprocal, asymmetrical and multi-directional exchange of care' (2013: 25), as well as 'family care regimes' (Kilkey & Merla, 2014) to denote the micro-sociologies of family arrangements. Rather revealing of wider social organisation is the fact that counter to such positive accounts of transnational family care in the international literature, there are compelling research contributions that highlight the challenges and difficulties encountered as fundamentally linked to transnational care and that strong transnational family ties are not necessarily the only or an inevitable outcome of transnational migration (e.g. Schröder-Butterfill & Newman, 2019; Schröder-Butterfill & Fithry, 2014).

Transnational family relations and the circulations of care are social practices that incorporate performativities of intimacy and affectivity. Such relations involve a number of challenges and their dynamics reveal the interplay of migrant agency and wider institutional structures. In the box content we focus on a few core empirical case studies and link their central findings to the wider literature to draw some wider reflections on interdisciplinary research on migration and family studies. While there are numerous empirical examples to draw on from the proliferation of research over the last 25 years on transnationalism and more recently on transnational families (Glick Schiller et al., 1992; Faist, 2000; Levitt, 2001; Chee, 2005; Parreñas, 2005; Carling et al., 2012) there is still much to highlight in focusing on a few selective case studies and drawing some key insights with a lens to gender and migration. In selecting case studies to illuminate some core insights regarding transnational families, we are also guided by the call to set aside methodological nationalism in providing a more global perspective on how migration driven social theories can explain global, national and local phenomena in a mutually constitutive way (Glick Schiller, 2009). We thus see both families and transnational family processes as components of transnational dynamics whereupon nation-states are core in shaping their relations but are also arenas of power dynamics (Glick Schiller, 2005).

For one, the large volume of literature on transnational families has focused almost exclusively on migrant transnational mothering, and, less so on fathers and fathering (Kilkey et al., 2014) (see Chap. 3), and, even less so from a queer perspective which reaffirms normative universalisations of gendered scripted lives of domesticity in heteronormative framings (Manalansan, 2006; Kosnick, 2011). Recent studies on black, migrant and gay/lesbian families (Moore, 2011; Capps & Fix, 2012) are seminal in extending issues of relationality, connectedness and intimacy by challenging heteronormative paradigms and introducing intersectional diversities and complexities in family lives.

4.5   Conclusion

Though previously understudied, movement due to familial and intimate reasons has grown enormously, especially as a result of the interest in how families live separated across space and time, the changing gendered and generational structures of the family, and how economic and emotional resources circulate between family and kin members to ensure their social reproduction. More recently, comparative research, both qualitative and quantitative, has yielded a better understanding of differences between countries and regions. As we have also seen in this chapter, a number of new perspectives have been developed to include men and masculinities in familial movements, tasks such as transnational parenting and to question the heteronormativity of assumptions. Increasingly same sex couples have been included within immigration regulations for family migrations, especially in Europe and other major societies of immigration.

References

Ambrosini, M., Bonizzoni, P., & Triandafyllidou, A. (2014). Family migration in southern Europe: Integration challenges and transnational dynamics: An introduction. *International Review of Sociology, 24*(3), 367–373.

Anthias, F. (2008). Thinking through the lens of translocational positionality: An intersectionality frame for understanding identity and belonging. *Translocations: Migration and Social Change., 4*(1), 5–20.

Anthias, F. (2009). Translocational belonging, identity and generation: Questions and problems in migration and ethnic studies. *Finnish Journal of Ethnicity and Migration, 4*(1), 6–15.

Anthias, F. (2012). Transnational mobilities, migration research and intersectionality. *Nordic Journal of Migration Research, 2*(2), 102–110.

Baldassar, L. (2007). Transnational families and aged care: The mobility of care and the migrancy of ageing. *Journal of Ethnic and Migration Studies, 33*(2), 275–297.

Baldassar, L. (2014). Too sick to move: Distant "crisis" care in transnational families. *International Review of Sociology, 24*(3), 391–405.

Baldassar, L., & Merla, L. (2013). *Transnational families, migration and the circulation of care: Understanding mobility and absence in family life*. Routledge.

Baldassar, L., Nedelcu, M., Merla, L., & Wilding, R. (2016). ICT-based co-presence in transnational families and communities: Challenging the premise of face-to-face proximity in sustaining relationships. *Global Networks, 16*(2), 133–144. https://doi.org/10.1111/glob.12108

Bélanger, D., & Silvey, R. (2019). An im/mobility turn: Power geometries of care and migration. *Journal of Ethnic and Migration Studies*. https://doi.org/10.1080/1369183X.2019.1592396

Berg, M. L. (2011). *Diasporic generation: Memory, politics and nation among Cubans in Spain*. Berghahn Books.

Bloch, A., & Hirsch, S. (2018). Inter-generational transnationalism: The impact of refugee backgrounds on second generation. *Comparative Migration Studies, 6*, 30. https://doi.org/10.1186/s40878-018-0096-0

Boccagni, P. (2012). Revisiting the "transnational" in migration studies: A sociological understanding. *Revue Européenne des Migrations Internationales, 28*(1). https://doi.org/10.4000/remi.5744

Boyd, M. (1989). Family and personal networks in international migration: Recent developments and new agendas. *International Migration Review, 23*(3), 638–670.

Browne, K., Banerjea, N., McGlynn, S. B., Bakshi, L., Banerjee, R., & Biswas, R. (2017). Towards transnational feminist queer methodologies. *Gender, Place & Culture, 24*(10), 1376–1397.

Bryceson, F. D. (2019). Transnational families negotiating migration and care life cycles across nation-state borders. *Journal of Ethnic and Migration Studies*. https://doi.org/10.1080/1369183X.2018.1547017

Bryceson, D., & Vuorela, U. (Eds.). (2002). *The transnational family: New European frontiers and global networks*. Berg Press.

Caarls, K., Haagsman, K., Kraus, E. K., & Mazzucato, V. (2018). African transnational families: Cross-country and gendered comparisons. *Population Space and Place, 24*(2), e2162.

Capps, R., & Fix, M. (Eds.). (2012). *Young children of black immigrants in America: Changing flows, changing faces*. Migration Policy Institute.

Carling, J., & Pettersen, S. V. (2014). Return migration intentions in the integration-transnational matrix. *International Migration, 52*(6), 13–30.

Carling, J., Menjívar, C., & Schmalzbauer, L. (2012). Central themes in the study of transnational parenthood. *Journal of Ethnic and Migration Studies, 38*(2), 191–217.

Cebotari, V., Mazzucato, V., & Siegel, M. (2017). Gendered perceptions of migration among Ghanaian children in transnational care. *Child Indicators Research, 10*(4), 971–993.

Charsley, K. (Ed.). (2012). *Transnational marriage: New perspectives from Europe and beyond*. Routledge.

Charsley, K., Bolognani, M., & Spencer, S. (2016). Marriage migration and integration: Interrogating assumptions in academic and policy debates. *Ethnicities*. https://doi.org/10.1177/1468796816677329

Chattopadhyay, S. (2018). Borders re/make bodies and bodies are made to make borders: Storying migrant trajectories. *ACME: An International Journal for Critical Geographies, 18*(1), 149–172.

Chauvin, S., Salcedo Robledo, M., Koren, T., & Illidge, J. (2021). Class, mobility and inequality in the lives of same-sex couples with mixed legal statuses. *Journal of Ethnic and Migration Studies, 47*(2), 430–446.

Chee, M. (2005). *Taiwanese American transnational families*. Routledge.

Christou, A., & Janta, H. (2019). The significance of things: Objects, emotions and cultural production in migrant Women's return visits home. *The Sociological Review, 67*(3), 654–671.

Christou, A., & King, R. (2015). *Counter-diaspora: The Greek second generation returns 'home'*. Harvard University Press.

Christou, A., & Michail, D. (2015). Migrating motherhood and gendering exile: Eastern European women narrate migrancy and homing. *Women's Studies International Forum, 52*, 71–81.

D'Aoust, A.-M. (2013). In the name of love: Marriage migration, governmentality, and technologies of love. *International Political Sociology, 7*(3), 258–274.

De Valk, H. A., & Medrano, J. D. (2014). Guest editorial on meeting and mating across borders: Union formation in the European Union single market. *Population, Space and Place, 20*(2), 103–109.

Devenney, K. (2020). 'My own blood': Family relationships of unaccompanied asylum-seeking young people in the UK. *Families, Relationships and Societies., 9*(2), 183–199.

European Migration Network. (2017). *Family reunification of third country nationals in the EU plus Norway: National practices*.

Eurostat. (2020). *Statistics on migration to Europe*. https://ec.europa.eu/info/strategy/priorities-2019-2024/promoting-our-european-way-life/statistics-migrationeurope_en

Faist, T. (2000). *The volume and dynamics of international migration and transnational social spaces*. OUP.

Fesenmyer, L. E. (2014). Transnational families: Towards emotion. In B. Anderson & M. Keith (Eds.), *Migration: A COMPAS anthology*. COMPAS.

Fournier, C., Hamelin Brabant, L., Dupéré, S., & Chamberland, L. (2018). Lesbian and gay immigrants' post-migration experiences: An integrative literature review. *Journal of Immigrant & Refugee Studies, 16*(3), 331–350.

Fresnoza-Flot, A., & Ricordeau, G. (Eds.). (2017). *International marriages and marital citizenship. Southeast Asian women on the move*. Routledge.

Gardner, K. (2012). Transnational migration and the study of children: An introduction. *Journal of Ethnic and Migration Studies, 38*(6), 889–912.

Gaspar, S. (2012). Global patterns of bi-national couples across five EU countries. *Sociologia, problemas e práticas, 70*, 71–89.

Gassmann, F., Siegel, M., Vanore, M., et al. (2018). Unpacking the relationship between parental migration and child well-being: Evidence from Moldova and Georgia. *Child Indicators Research, 11*, 423. https://doi.org/10.1007/s12187-017-9461-z

Glick Schiller, N. (2005). Transnational social fields and imperialism: Bring a theory of power to transnational studies. *Anthropological Theory, 5*(4), 439–461.

Glick Schiller, N. (2009). *A global perspective on transnational migration: Theorizing migration without methodological nationalism*. Centre on Migration, Policy and Society, Working Paper No. 67.

Glick Schiller, N., Basch, L., & Blanc-Szanton, C. (1992). Transnationalism: A new analytic framework for understanding migration. *Annals of the New York Academy of Sciences, 645*(1), 1–24.

Gorman-Murray, A. (2009). Intimate mobilities: Emotional embodiment and queer migration. *Social and Cultural Geography, 10*(4), 441–460.

Goulbourne, H., Reynolds, T., Solomos, J., & Zontini, E. (2010). *Transnational families: Ethnicities, identities and social capital*. Routledge.

Groes, C., & Fernandez, N. (2018). *Intimate mobilities and mobile intimacies*. Berghahn Books.

Haagsman, K. (2018). Do transnational child-raising arrangements affect job outcomes of migrant parents? Comparing Angolan parents in transnational and non-transnational families in the Netherlands. *Journal of Family Issues, 39*(6), 1498–1522.

Herz, M., & Lalander, P. (2017). Being alone or becoming lonely? The complexity of portraying unaccompanied children as being alone in Sweden. *Journal of Youth Studies, 20*(8), 1062–1076.

Ibrahim, F. (2018). Cross-border intimacies: Marriage, migration, and citizenship in western India. *Modern Asian Studies, 52*(5), 1664–1691.

Janta, H., & Christou, A. (2019). Hosting as social practice: Gendered insights into contemporary tourism mobilities. *Annals of Tourism Research, 74*, 167–176.

Johnson, C., Jones, R., Paasi, A., Amoore, L., Mountz, A., Salter, M., & Rumford, C. (2011). Intervention: Rethinking the border in border studies. *Political Geography, 30*, 60–69.

Jordan, L. P., Dito, B., Nobles, J., & Graham, E. (2018). Engaged parenting, gender, and children's time use in transnational families: An assessment spanning three global regions. *Population, Space and Place, 24*(7), E2159.

Juozeliūnienė, I., & Budginaitė, I. (2018). How transnational mothering is seen to be 'troubling': Contesting and reframing mothering. *Sociological Research Online, 23*(1), 262–281.

Kertzer, D. I. (1983). Generation as a sociological problem. *Annual Review of Sociology, 9*, 129–149.

Kilkey, M., & Merla, L. (2014). Situating transnational families' care-giving arrangements: The role of institutional contexts. *Global Networks, 14*(2), 210–229.

Kilkey, M., Plomien, A., & Perrons, D. (2014). Migrant men's fathering narratives, practices and projects in national and transnational spaces: Recent Polish male migrants to London. *International Migration, 52*(1), 178–191.

King, R., & Christou, A. (2010). Diaspora, migration and transnationalism: Insights from the study of second-generation 'returnees'. In R. Bauböck & T. Faist (Eds.), *Diaspora and transnationalism: Conceptual, theoretical and methodological challenges* (pp. 167–183). Amsterdam University Press.

King, R., & Vullnetari, J. (2006). Orphan pensioners and migrating grandparents: The impact of mass migration on older people in rural Albania. *Ageing and Society, 26*(5), 783–816.

King, R., Christou, A., & Teerling, J. (2011). 'We took a bath with the chickens': Memories of childhood visits to the homeland by second-generation Greek and Greek Cypriot 'returnees'. *Global Networks: A Journal of Transnational Affairs, 11*(1), 1–23.

Kofman, E. (2018). Family migration as a class matter. *International Migration, 56*(4), 33–46.

Kofman, E. (2019). 'Families on the move'. *Women's progress of the world's women 2019–2020*, Families in a Changing World, UN Women.

Kofman, E., & Raghuram, P. (2012). Women, migration, and care: Explorations of diversity and dynamism in the global south. *Social Politics, 19*(3), 408–432.

Kosnick, K. (2011). Sexuality and migration studies: The invisible, oxymoronic and heteronormative othering. In H. Lutz, L. Supik, & M. T. Herrera Vivar (Eds.), *Framing intersectionality: Debates on a multi-faceted concept in gender studies* (pp. 121–136). Ashgate.

Kõu, A., Mulder, C. H., & Bailey, A. (2017). 'For the sake of the family and future': The linked lives of highly skilled Indian migrants. *Journal of Ethnic and Migration Studies, 43*(16), 2788–2805.

Kufakurinani, U., Pasura, D., & McGregor, J. (2014). Transnational parenting and the emergence of 'diaspora orphans' in Zimbabwe. *African Diaspora, 7*(1), 114–138.

Lam, T., & Yeoh, B. (2018). Migrant mothers, left-behind fathers: The negotiation of gender subjectivities in Indonesia and the Philippines. *Gender, Place & Culture, 25*(1), 104–117.

Levitt, P. (2001). *The transnational villagers*. University of California Press.

Luibhéid, E. (2008). Queer/migration: An unruly body of scholarship. *GLQ: A Journal of Lesbian and Gay Studies, 14*(2), 169–190.

Madianou, M., & Miller, D. (2012). *Migration and new media, transnational families and polymedia*. Routledge.

Mai, N., & King, R. (2009). Love, sexuality and migration: Mapping the issue(s). *Mobilities, 4*(3), 295–307.

Manalansan, M. F. (2006). Queer intersections. Sexuality and gender in migration studies. *International Migration Review, 40*(1), 224–249.

Mand, K. (2015). Childhood, emotions and the labour of transnational families. *Discourse: Journal of Childhood and Adolescence Research, 1*, S.25–S.39.

McCarthy, J. R., & Edwards, R. (2011). Transnational families. In *The SAGE key concepts series: Key concepts in family studies* (pp. 188–190). SAGE.

Menchavez-Francisco, V. (2018). *Labor of care of Filipina migrants and transnational families in the digital age*. Illinois University Press.

Migali, S., & Natale, F. (2017). *The determinants of migration to the EU: Evidence from residence permits data*. European Union. https://ec.europa.eu/jrc/en/publication/determinants-migration-eu-evidence-residence-permits-data

Mole, R. (2016). Homonationalism: Resisting nationalist co-optation of sexual diversity. *Sexualities, 20*(5–6), 660–662.

Mole, R. (2018). Sexualities and queer migration research. *Sexualities, 21*(8), 1268–1270.

Moore, M. R. (2011). *Invisible families: Gay identities, relationships, and motherhood among black women*. University of California Press.

Moret, J., Andrikopoulos, A., & Dahinden, J. (2021). Contesting categories: Cross-border marriages from the perspectives of the state, spouses and researchers. *Journal of Ethnic and Migration Studies, 47*(2), 325–342.

Nedelcu, M. (2017). Transnational grandparenting in the digital age: Mediated co-presence and childcare in the case of Romanian migrants in Switzerland and Canada. *European Journal of Ageing, 14*, 375–383.

Ní Laoire, C., Carpena-Méndez, F., Tyrrell, N., & White, A. (2010). Introduction: Childhood and migration—Mobilities, homes and belongings. *Childhood, 17*(2), 155–162.

OECD. (2019). *International migration outlook*. OECD.

Parreñas, R. (2005). *Children of global migration: Transnational families and gendered woes*. University of Stanford Press.

Pasura, D., & Christou, A. (2018). Theorizing Black (African) transnational masculinities. *Men and Masculinities, 21*(4), 521–546.

Pellander, S. (2021). Buy me love: Entanglements of citizenship, income and emotions in regulating marriage migration. *Journal of Ethnic and Migration Studies, 47*(2), 464–479.

Portes, A., & Rumbaut, R. G. (2006). *Immigrant America: A portrait*. University of California Press.

Ramakrishnan, S. K. (2004). Second-generation immigrants? The "2.5 generation" in the United States. *Social Science Quarterly, 85*, 380–399.

Ryan, L., Sales, R., Tilki, M., & Siara, B. (2009). Family strategies and transnational migration: Recent Polish migrants in London. *Journal of Ethnic and Migration Studies, 35*(1), 61–77.

Safi, M. (2018). Varieties of transnationalism and its changing determinants across immigrant generations: Evidence from French data 1. *International Migration Review, 52*(3), 853–897.

Schröder-Butterfill, E., & Fithry, T. S. (2014). Care dependence in old age: preferences, practices and implications in two Indonesian communities. *Ageing & Society, 34*(3), 361–387.

Schröder-Butterfill, E., & Newman, J. (2019). Transnational families and the circulation of care: A Romanian-German case study. *Ageing & Society, 39*(1), 45–73.

Staver, A. (2015). Hard work for love. The economic drift in Norwegian family immigration and integration policies. *Journal of Family Issues, 36*(11), 1453–1471.

Vaughan-Williams, N., & Pisani, M. (2018). Migrating borders, bordering lives: Everyday geographies of ontological security and insecurity in Malta. *Social & Cultural Geography, 21*(5), 651–673.

Wimark, T. (2019). Homemaking and perpetual liminality among queer refugees. *Social & Cultural Geography*. https://doi.org/10.1080/14649365.2019.1619818

Yeoh, B. (2016). Migration and gender politics in Southeast Asia. *Migration, Mobility, & Displacement, 2*(1), 74–88.

Yuval-Davis, N., Wemyss, G., & Cassidy, K. (2018). Everyday bordering, belonging and the reorientation of British immigration legislation. *Sociology, 52*(2), 228–244.

Zeitlyn, B. (2012). Maintaining transnational social fields, the role of visits to Bangladesh for British Bangladeshi children. *Journal of Ethnic and Migration Studies, 38*(6), 953–968.

Zontini, E. (2010). *Transnational families, migration and gender: Moroccan and Filipino women in Bologna and Barcelona*. Berghan Book.

Zontini, E. (2015). Growing old in a transnational social field: Belonging, mobility and identity among Italian migrants. *Ethnic and Racial Studies, 38*(2), 326–341.

Chapter 5
Gendering Asylum

By the end of 2019, 79.5 million people of concern (refugees and internally displaced) around the world had been forced from their home countries. It represents over three times the number of people of concern compared to the figure at the beginning of the twenty-first century. The major development since the peak in asylum applications in 2015 in Europe has been the large-scale emigration of Venezuelans, who as of 2019 are now among the top three nationalities in Europe, especially in Spain, and the outflow from Afghanistan since the takeover by the Taliban in August 2021. On the other hand, Covid-19 has led to a significant reduction in applicants in 2020, especially among Colombians and Venezuelans arriving by air (EASO, 2021).

Women comprise the majority of those escaping generalized conflict, but only a minority of those who manage to seek asylum in the Global North due to the fact that moving long distances requires considerable resources and frequently necessitates the use of smugglers (Damir-Geilsdorf & Sabra, 2018). In many of the countries with large numbers of populations of concern, such as Colombia, DR Congo, South Sudan and Syria, women form the majority or almost half of the displaced population (UNHCR Statistical Yearbook, 2016, table 13). For example, Syria, which had become the largest refugee producing country, had an estimated 6.5 million Syrian citizens internally displaced and more than 4.8 million in neighbouring countries by the end of 2016. Women form the majority of the internally displaced in Syria itself and about half in neighbouring countries such as Jordan, Lebanon and Turkey (Freedman et al., 2017; Williams et al., 2020).

It was only in the 1980s that concerns about women in forced migration came to the fore among academics and international organisations (Indra, 1999). A gender approach, she noted, was still in its infancy in the 1990s, and would require more attention being paid to situationally specific and in-depth knowledge of women and men forced migrants, including the class, ethnic, national and transnational systems of which they are part (ibid: 21). Yet, as with other forms of migration, international statistics on the gender breakdown of refugee populations was for a long time not available, leading to the erroneous idea that the majority of assisted refugees in the Global South were women and children (Zlotnik, 2003). Though more statistics have

A. Christou, E. Kofman, *Gender and Migration*, IMISCOE Research Series,
https://doi.org/10.1007/978-3-030-91971-9_5

become available since 1998, these are often not collected systematically for each stage of the asylum process and the different outcomes (Belloni et al., 2018; Kofman, 2019).

In this chapter we firstly outline the growing attention paid to gendered aspects of forced migrations in the 1990s in the Global South (Hyndman, 2010; Fiddian-Qasmiyeh, 2014) and the ways in which such gendered movements have been represented. Whilst it was men who reached the Global North, far fewer women were able to submit claims for asylum and thereby obtain refugee status. According to the 1951 Refugee Convention a refugee is defined "as a person who owing to a well-founded fear of being persecuted for reasons of race, religion, nationality, membership of a particular social group or political opinion, is outside the country of his nationality and is unable or, owing to such fear, is unwilling to avail himself of the protection of that country; or who, not having a nationality and being outside the country of his former habitual residence as a result of such events, is unable or, owing to such fear, is unwilling to return to it". Feminist scholars drew attention to the fact that the Convention failed to incorporate gender-related persecution and suggested ways in which such considerations could be incorporated within the limits of the Convention (Crawley, 1999; Macklin, 1999).

Although following the disintegration of Yugoslavia in the first half of the 1990s, Europe had known large-scale displacement from East to West with a peak of 700,000 asylum seekers in 1992, the 2015 influx brought in twice the number of persons as in the earlier period. The high level reflected both recent conflicts in Syria as well as protracted conflicts in Afghanistan, Eritrea and Somalia. Furthermore, in the past few years a large-scale exodus of over four million persons from Venezuela have also sought refuge, largely in the Americas as well as Spain.

In the second section we highlight how the large-scale flows of 2015, labelled as a migration or refugee crisis by politicians and the media, were characterised as one of the most significant post-war developments. This time, the large-scale displacement into Europe, as opposed to refugees located in the Global South, intensified some of the ongoing critical discussions around gender and refugee issues. These included the need for more disaggregated data, not just by gender but also in relation to other social divisions, and greater knowledge about the gendered experiences of border crossings and journeys (Holloway et al., 2019; Pruitt et al., 2018). One of the contentious issues has been the depiction of refugee women as victims and vulnerable (Belloni et al., 2018; Kofman, 2019; Parrs, 2018), the critique of the concept of vulnerability in humanitarian policies (Sozer, 2020; Turner, 2019b) and its implications for women, men and unaccompanied minors. Another emerging topic in academic and policy circles has been the treatment of sexuality-based asylum claims (Arbel et al., 2014) and the reception experiences of LGBTQI individuals (Henley, 2020).

5.1 Emergence of Gendered Perspectives on Forced Migration

The study of gender and refugees was slow to take off and remained fragmented for some time until the 1990s (Hyndman, 2010; Indra, 1999). Until gendered disaggregated data became available, there was an assumption that displaced persons were overwhelmingly female but the figures released in 1998 demonstrated that women only amounted to just less than half of assisted refugees in Africa (Zlotnik, 2003). Even now disaggregated data by sex are available for only 46% of the total UNHCR population of concern.

It has been argued that gendered imagery fundamentally shifted the representation of the refugee from a heroic European male to a depoliticised Third world mother and child or the womananddhild (Enloe, 1989) depicted as victims of generalised violence and poverty. This made it easier to attract funding for humanitarian assistance in the South through which the state played out its paternalistic role of saviour and protector (Johnson, 2011). Although UNHCR (1991) had adopted its policy on refugee women in 1990s, this tended to focus on women in their reproductive and domestic roles as defined in the World Conference on Women in Nairobi in 1985 rather than gender equality. Instead, it has been argued that traditional gendered images of the vulnerable and dependent female in need of protection have dominated refugee policies (Baines, 2004). Hyndman and Giles (2017) also argue that those who stay in the Global South are viewed positively as genuine, immobile, depoliticised, and feminised, while those on the move, in particular if aiming to reach the Global North, are perceived in negative terms as potential liabilities and/or security threats, which, as we shall see, is particularly associated with young refugee men.

Another area of critique and activism involved the 1951 Refugee Convention which, though supposedly neutral, was formulated around male norms and did not acknowledge gendered experiences of persecution (Crawley & Lester, 2004). It privileged the persecution of the actor in the public sphere in contrast to experiences in the private sphere of the family and home which might include familial and domestic violence, rape, and female genital mutilation. UNHCR recognized that 'historically, the refugee definition has been interpreted through a framework of male experiences, which has meant that many claims of women and of homosexuals have gone unrecognised' (UNHCR, 2002: n. 1), but suggested such recognition should be done through gender sensitive guidelines. The list of grounds of discrimination in the Convention were race, religion, nationality, political opinion, or membership in a particular social group, but did not include sex or gender. While gender sensitive guidelines had been passed in Canada (1993), United States (1995) and Australia (1996) (Macklin, 1999), progress was much slower in Europe (Ali et al., 2012). Although the guidelines have no status at the level of international law, they do spell out what it means to take into account gender-related persecution and issues of evidence and credibility assessment in refugee determination (Arbel et al., 2014). Importantly they enable women to make claims on the basis of persecution by private actors and in private spaces and have paved the way for sexuality-related claims. UNHCR argued that persecution based on gender, gender identity, and sexual orientation all stem from a common source, that of non-conformity to rigidly defined gender roles and gender norms (UNHCR, 2002, 2012). In its Guidelines, the UNHCR stated that "female applicants may be subjected to the same forms of harm as male applicants but they may also face forms of persecution specific to their sex, such as sexual violence, dowry-related violence, female genital mutilation, domestic violence and trafficking".

Yet the way in which gendered and sexual persecution should be recognised in the Convention also gave rise to debate between divergent ways of responding to its absence. The first is to classify women and LGBTQI claims within the remit of membership of particular social groups, the preferred option favoured by the UNHCR and the European Union for claims on the basis of gender and sexual persecution (Arbel et al., 2014), and which has prevailed in virtually all European countries (Ali et al., 2012); the second is to recognize these forms of persecution through the nexus of political opinion, nationality or religious identity (Crawley, 1999, 2021).

Within the EU in the first decade of the century, an average of about a third of female asylum seekers masked substantial differences from just over 10% to just under 50% in Poland. Another difference was whether disaggregated data were published (Ali et al., 2012). In addition the extent to which women asylum seekers were granted the more secure refugee status as opposed to subsidiary protection also differed markedly. For example, in Sweden the higher percentage given subsidiary protection statuses, a lower level than the Convention refugee status, reflected the fact that membership of a particular social group, to which gender-related persecution was aligned, was granted a lower level of protection. Only Romania included gender as a ground of persecution, while other countries interpreted gender-related persecution as falling in the category of a particular social group (Ali et al., 2012). How each form of persecution was interpreted and recognised for purposes of asylum determination also varied between countries. Following the gender guidelines, issues of sexuality began to draw attention.

> **Box: Sexuality and Grounds for Persecution**
> The subsequent decade saw quite a lot of activity around the recognition of sexuality as grounds of persecution with the UNHCR (2008) publishing its guidelines on claims relating to sexual orientation and gender identity in 2008 and the European Union recognizing sexual orientation as a cause of persecution in Article 10 of the EU Asylum Qualification Directive (2011) (Lewis & Naples, 2014). As with gender, this form of persecution was slotted into the category of particular social groups even though there might be other relevant grounds, such as cases where activism around LGBT issues might be seen as being in opposition to prevailing political or religious views and practices (Fiddian-Qasmiyeh, 2014). However unlike gender-related persecution, a comparative report on fleeing homophobia (Jansen & Spijkerboer, 2011) found there was virtually no data collected and a substantial disparity on how claims were dealt with.

Despite advances on gender-sensitive guidelines in some countries in the 1990s, there were still few comprehensive studies of women asylum seekers and refugees (Bloch et al., 2000). Indeed it was surprising that a large-scale intra-European forced migration phenomenon received so little attention from a gendered perspective. As previously noted, the disintegration of Yugoslavia and ethnic cleansing in Bosnia and Herzegovina (BiH) in the first half of the 1990s led to massive outflows to countries such as Australia and the United States as well as to other European states. Germany, which received the largest number, only gave a temporary status (tolerated or humanitarian). Thus at the end of the conflict in December 1995, the vast majority in Germany were repatriated to BiH. Of the 320,000 from BiH in Germany in 1992–5, only 22,000 remained in 2005. Unlike the minority of women applying for asylum in Europe, estimated to be about 29% in the mid 1990s (Crawley & Lester, 2004), Bosnian emigration according to statistics for BiH emigrants in OECD countries comprised 51.4% women. There were generally more women than men among refugees in the 15–29 years group, and especially for those 20–24 years old. It seems surprising that this large-scale displacement has generated few publications (Franz, 2003; Kačapor-Džihić & Oruč, 2012; Muftić & Bouffard, 2008), thus maintaining a binary portrayal of a female refugee majority in the South and a minority of female asylum claimants who have manged to cross borders into Europe (Johnson, 2011) to prevail.

In terms of conflict-inducing displacement, the role of sexual violence in generating displacement gained credence through the systematic use of rape in Bosnia in the 1990s and later Rwanda. Article 7(1g) of the Rome Statute of the International Criminal Court, in force since 2002, includes 'Rape, sexual slavery, enforced prostitution, forced pregnancy, enforced sterilization, or any other form of sexual violence of comparable gravity' as crimes against humanity when they are committed in a widespread or systematic way. Subsequently it has been argued that the practice of gender-based violence against boys and men in war and post-conflict situations has also to be recognised rather than treated simply as degrading treatment (Carpenter, 2006).

During the 1990s and the increase in refugee numbers in Europe, states proliferated legal statuses and associated rights to work and generated a stratified system of social protection (Kofman, 2002; Morris, 2002). However by the first decade of this century, the number of asylum seekers had been much reduced, in part due to increasingly restrictive bordering measures. Soon this would change as a series of new and protracted conflicts in the Middle East, North Africa and Africa combined to produce the unprecedented numbers seeking asylum and refugee status in Europe (see Table 5.1). The percentage of women hovered around a third.

In the next section we turn our attention to issues concerning the gendered aspects of the 2015 influx and an increasing percentage of female applicants (38.1% of first applicants out of a total of 612,700 in 2019 and 36.1% out of 416,600 in 2020 for EU-27 countries).

Table 5.1 Asylum seekers in the European Union (by sex and year of application)

Year	Female applicants	Male applicants	Percentage of women
2008	72.745	183.331	28
2009	93.950	203.075	32
2010	97.170	187.650	34
2011	106.355	235.315	31
2012	126.240	247.205	34
2013	150.760	307.710	33
2014	195.885	466.100	30
2015	384.995	1.006.160	28

Source: EUROSTAT online database

5.2   Displacement to Europe

Ever since the 1990s and the end of the Cold War, European states and then the European Union had been tightening rights to access not just into the territory and residence through its bordering processes, both externally and though everyday practices within states. Bordering is not only about who moves but also who controls the movement and under what conditions (de Genova, 2017; Yuval-Davis et al., 2019). In doing so, it filters and stratifies according to categories of nationality, education, age and gender, and who is perceived as likely to belong to and integrate into a modern society (see Chap. 6). Both within the EU and in individual states, an arsenal of policies contributed to an intensification of classifications, categories of eligibility and special spaces designated for asylum seekers. These included policies designating which country was responsible for asylum claims, for example, as in the Dublin Regulation, originally implemented in 1997 and changed several times since then with the aim of reducing 'asylum shopping'. Some countries were designated as safe and from which claims for asylum claims were set aside; others were recognised as places of conflict and thereby valid sites for claims. This gave rise to very different rates of recognition of claims from the Syrians with at the end of 2016 the highest rate of recognition of 98% (refugee, subsidiary protection, humanitarian) and Eritreans with 93% whilst others, particularly from sub-Saharan Africa, such as Nigerians, and Pakistan had had very high rates of rejection of over 80% (Eurostat Asylum Statistics). Special spaces or hotspots (D'Angelo, 2019) effectively serving as spaces of detention, were also created in several sites in Sicily as from the end of 2015 and then on the Greek Islands to filter the 'genuine' asylum seeker, often reduced to nationality, from the economic migrant (Crawley & Skleparis, 2018). However as D'Angelo (2019) cautions, for Italy these were simply factories manufacturing illegality since applicants were rarely repatriated but left to remain undocumented and without rights. In part the nationality classification is inflected by humanitarian principles influenced increasingly by categories derived from the concept of vulnerability (Peroni & Timmer, 2013). Those to whom the label of 'vulnerable' is affixed are given priority in border crossings and access to resources. As we shall show, women disproportionately fall into these categories (pregnant, single parents) in contrast to men who are more likely to be seen as threatening or able to look after themselves (Kofman, 2019).

Whereas the early flows of asylum seekers in Europe were predominantly men, the share of women rose as from the end of 2015 and until the EU-Turkey deal in March 2016 which closed off this particular route. The visible dominance of men generated negative, though sometimes contradictory, comments on men who had fled to Europe (Herz, 2019; Scheibelhofer, 2017, 2019). Analysis of social media portrayed them as threatening to society and women in general or as cowards unwilling to fight, having left women and children behind to fend for themselves (Helms, 2015; Pruitt et al., 2018; Rettberg & Gajjala, 2016). So men emerged as potential terrorists and security threats, which would be reinforced by subsequent events in the Paris bombings of November 2015, the incident on the Thalys train from Brussels to Paris and the bombings in Brussels in March 2016, for which it was suggested that some of the perpetrators had returned among refugees. This and security fears also had repercussions on resettlement programmes resulting in the Canadian government excluding single, except gay, men from its Syrian resettlement programme (Kingsley, 2015). Others questioned taking in such male-dominated populations, especially among unaccompanied minors who would soon transition to adulthood, and pose a threat to Europe's gender equal societies (Hudson, 2016; Pruitt et al., 2018). A majority of tweets on #refugeesNOTwelcome invoked gender-based arguments or imagery against immigration or refugee settlement and explicitly linked the arrival of refugees to gender-based violence or the subjugation of women (Ingulfsen, 2016; Kreis, 2017).

For the Mediterranean sea crossings as a whole in 2015, 58% of the 1,015,078 were men, 17% women and 25% children though masking quite divergent patterns. However, by November 2015, a shift to an increasing number of women, including single and pregnant women, and children was being reported for the Greek route (UNHCR et al., 2016), a 10% increase since May 2015. The percentage of women among asylum applicants in Germany had increased from 21% in 2015 to 32% in 2016 (Damir-Geilsdorf & Sabra, 2018). It is likely that one of the reasons this happened was family separation at different stages of the process and the difficulties in re-joining family members (Costello et al., 2017; Damir-Geilsdorf & Sabra, 2018). The slow process of family reunification in countries of origin with very long waiting times to obtain papers meant that some left without waiting for official permission (Squire et al., 2017). Getting out of Greece to join family members could also take time, especially once Germany and Sweden put a brake on family reunification from the end of 2015 and 2016 for those with subsidiary protection (see Chap. 4). Yet in Italy, with much lower numbers in 2015, and which had very different source countries primarily from Africa, the percentage of female migrants remained low (Table 5.2).

The percentage of women also varied considerably between nationalities. In Greece, at the end of 2015 the nationalities with highest shares of women were Syrians (43%), Afghanis (29%) and Iraqis (12%). In Italy, it was Nigerians who had the highest rate of women (25%) followed by Eritreans (22%) and Somalis (21%).

Refugee women are a sizeable and growing group. According to data from Eurostat, about half a million women obtained international protection in Europe

Table 5.2 Percentage of men, women and children among arrivals in 2015 and 2016

	2015		2016 (Jan–Sept)	
	Greece	Italy	Greece	Italy
Men	55	75	41	61
Women	17	14	21	12
Children	28	11	38	27[a]

Source: UNHCR
[a]In Italy unlike in Greece, there were large numbers of unaccompanied minors so that the 27% children was made up of 14% accompanied and 13% unaccompanied children

since 2015, of whom 300,000 are in Germany. The presence of refugee women is also expected to rise further through family reunification (see Chap. 4), as the majority of spouses concerned are women. The greatest gender differences were observed for asylum applicants who were 14–17 or 18–34 years old, where 67.9% and 69.0%, respectively, of first-time applicants were male, with this share dropping back to 58.0% for the age group 35–64 years (Eurostat, May 2020). Unaccompanied minors remain overwhelmingly male but there is very little gender disaggregated data of children as if they were gender neutral (Kofman, 2019).

However despite the growing availability of data on gender breakdown together with age, disaggregation does not reveal the heterogeneity of asylum seekers characteristics, with whom refugees have travelled and their aspirations. Disability (Rohwerder, 2018) and age too are highly relevant in the way asylum seekers and refugees experience their journeys and settlement. Yet we have little information or data on disability, despite the UNHCR having recognised refugees with disabilities as a group to whom it had obligations (Fiske & Giotis, 2021). It can be seen as reflecting a focus in forced migration on heternonormative productive bodies (Pisani & Grech, 2015).

Disaggregating data would allow us to gain a better understanding of the politics of gendered mobilities and unequal access to mobility (Uteng & Cresswell, 2008). For this we need to turn to smaller surveys and qualitative research based on ethnographies, interviews and films. A number of surveys were conducted during the peak of the Mediterranean crossings. For example, in the first wave of the survey (March–July 2016) for the project *EVI-MED: Constructing an evidence base of contemporary Mediterranean migrations* (Blitz et al., 2017), women in Greece were far more likely than men to be divorced or widowed (9 women compared to 2 men), while 9 were single so that a third were, therefore, without a husband. The majority had children with them in Greece, few (2) had left children behind in the country of origin with 9 of them having children elsewhere. Few women (5) had travelled alone compared to men (26). UNHCR (2016a, b) also conducted interviews at the beginning of 2016 with Syrians and Afghans on the islands (Lesvos, Chios, Samos and Leros). Of the 524 Syrians interviewed, 23% were women of whom 2% were pregnant and 2% lactating. 80% had travelled with close family, 10% with extended family, 7% with friends and colleagues and only 11% were alone. 18% of respondents were part of a single male-headed household and 19% a female-headed household. 7% had left behind a spouse, 40% a parent and 13% children.

In Italy, as previously noted, there were far fewer women. The EVI-MED survey comprising 202 individuals (March–June 2016) indicated that of the 23 women surveyed, 14 were single and three widowed. 12 did not have any children and, of the 11 who did, only 3 were living with them in Sicily. Although fewer women had travelled alone (60%) compared to men (76%), this is considerably higher than for women in Greece.

We therefore need to delve more deeply to capture the experiences of women, men and children (UNICEF, 2020) as they cross international borders under inhospitable conditions and understand the relationship between the harm, distress and violence many are subjected to as well as the agency they deploy (Grotti et al., 2018; Holloway et al., 2019; Oxfam, 2016). It is particularly important we do so in order not to insert their stories into prevailing stereotypes of asylum seekers and migrants. As we have previously noted, sex work is frequently coupled with sex trafficking (Chap. 3), especially for certain nationalities, such as Nigerian women in Italy, who are rendered as pure victims without any agency. Most had traversed Libya, a highly dangerous and violent country where many individuals had experienced serious harm of sequestration, forced labour, kidnapping and sexual violence from a variety of sources. It is not easy to distinguish sex trafficking from using transactional sex to undertake a journey (Crawley et al., 2016; Hodal, 2020). Nigerian women in particular are closely associated with sex work (Plambech, 2017; Rigo, 2017).

Taking away agency has also been problematic in the increasing application of vulnerability criteria in relation to particular categories of asylum seekers and refugees, often pushing them to perform vulnerability in order to be prioritised for allocation of resources. For example, November 2015, UNHCR financed NGOs in Greece to offer housing, either in hotels or apartments, to eligible asylum seekers, such as those enrolled in the EU Relocation Scheme, Dublin family-reunification candidates, and, since 2016, "vulnerable" applicants. Others may self-vulnerabilise in order to gain access to resettlement schemes to wealthier countries (Parrs, 2018). The name of the UK Syrian Vulnerable Persons Scheme reflects this in its stated preference for families, thus excluding single men (Turner, 2017).

5.3 Vulnerability

The concept of vulnerability has emerged in the past two decades in political, social and legal theory (Fineman, 2008; Turner, 2006), in ethics and public health (Luna, 2019) and in humanitarian policies (Heidbrink, 2021; Sozer, 2020; Turner, 2019a, b). Martha Albertson Fineman starts from a critique of the liberal notion of the autonomous individual which she argued should be replaced by the vulnerable subject 'describing a universal, inevitable and enduring aspect of the human condition that must be at the heart of our concept of social and state responsibility' (2008: 8). For her the condition of vulnerability should be understood as stemming from our embodiment which carries the possibility of harm, injury and misfortune in the past, present and future, and which may render us more dependent over the life course.

Thus vulnerability represents connectivity between individuals and applies to everyone, and not simply to designated groups, as in the approach to vulnerability adopted in the European Human Court of Human Rights case law (Al Tamini, 2015; Peroni & Timmer, 2013) and in UNHCR humanitarian interventions (Sozer, 2020; UNHCR, 2014).

A number of critiques have been levelled at the application of vulnerability and its classification of vulnerable individuals and groups. Initially asylum seekers as a group (ECRE, 2017) were considered to be vulnerable and in need of special protection as in the case of M.S.S. v Belgium and Greece (ECtHR [GC] 21 January 2011, no. 30696/09 (M.S.S. v. Belgium and Greece) in which the systematic deficiencies of the Greek asylum system, such as a lack of reception centres, inability to access the labour market, lengthy procedures in having asylum requests examined and the traumas asylum seekers had been through during the process of migration could be said to contribute to 'the institutional production of vulnerability of asylum seekers in Greece' (Peroni & Timmer, 2013: 1069). Yet, as the number of asylum seekers increased, so too has the label of vulnerability been restricted to particular groups (ECRE, 2017) designated by the European Union in its Qualification Directive (2011) and the UNHCR (2013).

However, we know little about the recipients of prioritisation and what the impact of being designated vulnerable has had on their lives. A partial exception was the pre-registration exercise in summer 2016 in Greece which provided a picture of the composition of the vulnerable population (see Table 5.3). The majority of adults were women due to the large numbers of those who were pregnant, had recently given birth or were single parents with children. Among the vulnerable categories listed, there seemed to be a tendency to privilege protection based on past harm, such as disability, torture, and exploitation, or those who care for or are dependent on

Table 5.3 Vulnerabilities by type and gender in Greece. Pre-registration June–July 2016

Category of vulnerability	Male	% of total male adults	Female	% of total female adults	Total no.
Single parents with minor children	104	15.3	627	38.4	731
Pregnant women/recently given birth	0	–	522	32.0	522
Incurable or serious diseases	174	26.6	174	10.7	348
Disability	209	30.8	104	6.4	213
Elderly	104	15.3	139	8.5	243
Post-traumatic disorder	39	5.7	39	2.4	78
Torture	39	5.7	10	0.6	49
Rape or serious exploitation	10	1.5	17	1.0	27
Total adults	679		1632		3481
Total adults and unaccompanied minors	1688		1841		3481

Source: Hellenic Republic, Ministry of Interior and UNHCR pre-registration data analysis 9 June-−30 July 2016

others, such as single parents with young children, or those who require additional support, for example, pregnant women, the elderly, the disabled and unaccompanied children. What also distinguishes most of these categories are that they are the most visible and easily identifiable, though those with mental health problems may not want to disclose this. This check list expedites the process of classification, as a hard-pressed doctor working with disembarking asylum seekers in Italy, stated (Heidbrink, 2021).

> **Box: Vulnerable Groups**
> However, an odd absence from this list are single women travelling on their own who have been identified in a number of reports (Women's Refugee Commission, 2016a, b) and their own stories of dangerous and threatening situations and gender- based violence from a variety of actors (Rigo, 2017). Transactional or survival sex was demanded to cross a border or advance on their journey, especially amongst those travelling from sub-Saharan Africa through Libya. Yet, the reliance on a labelling or listing approach to facilitate the governance of asylum and refugee management neglects those whose vulnerability derives primarily from their insecure situations.

Apart from the critique of this classification having become a listing exercise that fails to take into account a more contextual and situated concept of actual and potential harm, others have highlighted the absence of men from this conceptualisation (Turner, 2019a). Indeed the remit of organisations is often limited to assessment of risks faced by individuals though they have noted risks for boys and men arising from forced conscription and traumatic journeys (UNHCR et al., 2016). More comprehensive critiques have surfaced recently in relation to its close association with neo-liberal humanitarianism and rationing of resources with problematic consequences in its redistribution between refugees (Sozer, 2020). For Heidbrink (2021), it is a means that states and humanitarian actors deploy to govern contemporary mobility and restrict access to much reduced services in increasingly privatised welfare regimes. Turner's shift in position is quite instructive. Having argued for the inclusion of men as vulnerable subjects, which consequently would 'disrupt prevailing humanitarian understandings of refugeehood as a feminized subject position' (Turner, 2019a), he subsequently (Turner, 2019b: 17) forcefully argued that we do not need more studies of refugees' "vulnerabilities" or categories such as men and LGBT (Myrttinen et al., 2017) to be incorporated. Instead we need studies of refugees' lives that are grounded in their own concepts and understandings and do not force them into performing powerlessness in order to acquire vulnerability. And as others have also commented, vulnerability has reduced their subjectivity to this aspect and stripped them having any agentic qualities. Yet at the same time, others see the addition of categories such as LGBTI into EU Directives as a good solution to pushing governments to recognise claims made on this basis (FRA, 2017).

With women in particular, the focus on vulnerability has emphasised sexual and gender-based violence in their journeys crossing the Mediterranean and Europe and in reception facilities, especially in Germany and Sweden which received the largest number of asylum seekers (Bonewit & Shreeves, 2016; Honeyball, 2016; Women's Refugee Commission, 2016a, b). A number of scholars have critiqued the reduction of their situation to one of the official definitions of vulnerability (Belloni et al., 2018; Fiddian-Qasmiyeh, 2014; Freedman, 2015; Freedman et al., 2017) and the failure to take account of the complexity of their situations. Some have argued that the excessive emphasis of women's victimisation (Pittaway & Bartolomei, 2018), has rendered it difficult to develop appropriate measures for supporting them in transit and in the country of destination in relation to sexual and gender-based violence (Grotti et al., 2018; Ozcurumez et al., 2020).

So, too, have unaccompanied children been represented as quintessentially vulnerable (Heidbrink, 2021; Pruitt et al., 2018) without taking into account their aims and aspirations, especially for those escaping countries of protracted conflict without any sense of future or opportunities (Belloni, 2020). Their categorisation as vulnerable persons fixes them temporally into a particular status in their life course without taking into account their continuing vulnerability, especially as they confront the insecurity of their transition into adulthood (Heidbrink, 2021).

Whilst sexual orientation and gender identity are included in the Qualifications Directive 2011 (FRA, 2017) laying down which grounds are eligible for international protection, they are not enumerated in the Reception Directive 2013 as grounds of vulnerability though it has been argued that the grounds can be extended. Above all, many asylum seekers complain about the expectations that are expected of them to demonstrate that they are LGBTI, that is the credibility assessment. Across Europe, four in ten asylum seekers with such claims experience a 'culture of disbelief' that they had suffered or were at risk of persecution (Danisi et al., 2020; Henley, 2020). Sexuality and gender identity intersected with other reasons (country of origin, cultural background, demeanour, educational background, religion) in leading to their claims being rejected. In terms of reception facilities, there have over time been attempts to provide LGBTI asylum seekers, who are often harassed, with more secure accommodation, as has been the case in Germany (AIDA/ECRE, 2020).

5.4   Conclusion

In this chapter, we have traced the growing attention paid to gender and sexuality aspects of asylum and refugee flows, claims making and protection. No longer is it a matter of the analysis of the displaced being located some distance away in the Global South. It is important to adopt an historical perspective since as, we highlighted, the rapid growth of refugee claims had an earlier presence in the 1990s following the break-up and ethnic cleansing in the former Yugoslavia. Then, as now, the common response on the part of European states has been to stratify those making claims into diverse categories bearing different rights to remain and access legal and socio-economic rights in the Global North.

Today 20 plus years on, we see a humanitarian system in crisis where states and the European Union under neo-liberal governance and hostile environments towards migrants and refugees have left humanitarian and profit-making sectors to manage securitization of borders, including the filtering into categories. One of the means of processing asylum seekers into groups, between those to be settled with rights, those left in limbo and those to be deported, is the application of vulnerability in conjunction with nationality, often reflecting racialised notions of the other. As we have also explored, vulnerability has generated considerable critique, initially around which categories were included and excluded followed by a more radical questioning of the application of vulnerability altogether and the ways it has served to restrict access to protection and services. As Judith Butler (2016: 15) commented: 'Once groups are marked as 'vulnerable' within human rights discourse or legal regimes, those groups become reified as definitionally 'vulnerable', fixed in a political position of powerlessness and lack of agency. All the power belongs to the state and international institutions that are now supposed to offer them protection and advocacy'.

References

AIDA/ECRE. (2020). *Special reception needs of vulnerable groups*. Germany
Al Tamini, Y. (2015). *The protection of vulnerable groups and individuals by the European Court of Human Rights*. MSc Tilburg Law School.
Ali, H. C., Querton, C., & Soulard, E. (2012). *Gender related asylum claims in Europe. A comparative analysis of law, policies and practice focusing on women in nine EU Member States France, Belgium, Hungary, Italy, Malta, Romania, Spain, Sweden and the United Kingdom*. Directorate General for Internal Policies, Policy Department C: Citizens' Rights and Constitutional Affairs, Gender Equality.
Arbel, E., Dauvergne, C., & Millbank, J. (Eds.). (2014). *Gender in refugee law: From the margins to the Centre*. Routledge.
Baines, E. (2004). *Vulnerable bodies: Gender, the UN and the global refugee crisis*. Ashgate.
Belloni, M. (2020). Family project or individual choice? Exploring agency in young Eritreans' migration. *Journal of Ethnic and Migration Studies, 46*(2), 336–353.
Belloni, M., Pastore, F., & Timmerman, C. (2018). Women in Mediterranean asylum flows: Current scenario and ways forward. In C. Timmerman, M. L. Fonseca, L. v. Praag, & S. Pereira (Eds.), *Gender and migration: A gender-sensitive approach to migration dynamics* (pp. 195–216). University of Leuven Press.
Blitz, B., D'Angelo, A., Kofman, E., & Montagna, N. (2017). *Mapping refugee reception in the Mediterranean*. First Report of the EVI-MED Project. Middlesex University. www.mdx.ac.uk/evimed
Bloch, A., Galvin, T., & Harrell-Bond, B. (2000). Refugee women in Europe: Some aspects of the legal and policy dimensions. *International Migration, 38*(2), 169–188.
Bonewit, A., & Shreeves, R. (2016). *Reception of female refugees and asylum seekers in the EU: Case study Germany*. Directorate General for Internal Policies, Policy Department C: Citizens' Rights and Constitutional Affairs Women's Rights & Gender Equality.
Butler, J. (2016). Rethinking vulnerability and resistance. In J. Butler, Z. Gamberri, & L. Sabsay (Eds.), *Vulnerability in resistance* (pp. 12–27). Duke University Press.

Carpenter, C. (2006). Recognizing gender-based violence against civilian men and boys in conflict situations. *Security Dialogue, 37*(1), 83–103.

Costello, C., Gorenendijk, K., & Storgaard, L. (2017). *Realising the right to family reunification of refugees in Europe*. Refugee Studies Centre, University of Oxford.

Crawley, H. (1999). Women and refugee status: Beyond the public/private dichotomy in UK asylum policy. In D. Indra (Ed.), *Engendering forced migration* (Theory and practice) (pp. 308–333). Berghahn Books.

Crawley, H. (2021). Gender, 'refugee women' and the politics of protection. In C. Mora & N. Piper (Eds.), *The Palgrave handbook of gender and migration* (pp. 359–372). Palgrave Macmillan.

Crawley, H., & Lester, T. (2004). Comparative analysis of gender-related persecution in national asylum legislation and practice in Europe. *UNHCR Regional*.

Crawley, H., & Skleparis, D. (2018). Refugees, migrants, neither, both: Categorical fetishism and the politics of bounding in Europe's 'migration crisis'. *Journal of Ethnic and Racial Studies, 44*(1), 48–64.

Crawley, H., Düvell, F., Jones, K., McMahon, S., & Sigona, N. (2016). *Destination Europe? Understanding the dynamics and drivers of Mediterranean migration in 2015*. MEDMIG Final Report www.medmig.info/research-brief-destination-europe.pdf

D'Angelo, A. (2019). Italy: The 'illegality factory'? Theory and practice of refugees' reception in Sicily. *Journal of Ethnic and Migration Studies, 45*, 2213–2226.

Damir-Geilsdorf, S., & Sabra, M. (2018). *Disrupted families. The gender impact of family reunification policies on Syrian refugees in Germany*. UN Women Discussion Paper 23.

Danisi, C., Dustin, M., Ferreira, N., & Held, N. (2020). *Queering asylum in Europe: Legal and social experiences of seeking international protection on grounds of sexual orientation and gender identity*. Springer.

De Genova, N. (2017). Introduction. The borders of "Europe" and the European question. In N. De Genova (Ed.), *The borders of "Europe": Autonomy of migration, tactics of bordering* (pp. 1–35). Duke University Press.

Directive 2011/95/EU of the European Parliament and of the Council of 13 December 2011 on standards for the qualification of third-country nationals or stateless persons as beneficiaries of international protection, for a uniform status for refugees or for persons eligible for subsidiary protection, and for the content of the protection granted.

Directive 2013/33/EU of the European Parliament and of the Council of 26 June 2013 laying down standards for the reception of applicants for international protection. http://eur-lex.europa.eu/LexUriServ/LexUriServ.do?uri=OJ:L:2013:180:0096:0116:EN:PDF

EASO. (2021). *Asylum Trends*. https://easo.europa.eu/latest-asylum-trends

ECRE. (2017). *The concept of vulnerability in European asylum procedures*. Asylum Information Database European Council on Refugees and Exiles. http://www.asylumineurope.org/sites/default/files/shadowreports/aida_vulnerability_in_asylum_procedures.pdf

Enloe, C. (1989). *Beaches. Bananas and bases*. Pandora Press.

Eurostat. (2020). *Asylum Statistics*. https://ec.europa.eu/eurostat/statistics-explained/index.php/Asylum_statistics

Fiddian-Qasmiyeh, E. (2014). Gender and forced migration. In E. Fiddian-Qasmiyeh, G. Loescher, K. Long, & N. Sigona (Eds.), *The Oxford handbook of refugee and forced migration studies* (pp. 395–408). Oxford University Press.

Fineman, M. (2008). The vulnerable subject: Anchoring equality in the human condition. *Yale Journal of Law & Feminism, 20*(1), 1–23.

Fiske, L., & Giotis, C. (2021). Refugees, gender and disability: Examining interactions through refugee journeys. In C. Mora & N. Piper (Eds.), *The Palgrave handbook of gender and migration* (pp. 441–454). London/New York.

Franz, B. (2003). Transplanted or uprooted? Integration efforts of Bosnian refugees based upon gender, class and ethnic differences in New York City and Vienna. *The European Journal of Women's Studies, 10*(2), 135–157.

Freedman, J. (2015). *Gendering the international asylum and refugee debate* (2nd ed.). Palgrave Macmillan.

Freedman, J., Z. Kivlicim and K. Baklauciglu, K. (eds) (2017). A gendered approach to the Syrian refugee crisis.

Fundamental Rights Agency (FRA). (2017). *Current migration situation in the EU: Lesbian, gay, bisexual, transgender and intersex asylum seekers*. FRA.

Grotti, V., Malakasis, C., Quagliariello, C., & Sahraoui, N. (2018). Shifting vulnerabilities: Gender and reproductive care on the migrant trail to Europe. *Comparative Migration Studies, 6*, 23. https://doi.org/10.1186/s40878-018-0089-z

Heidbrink, L. (2021). Anatomy of a crisis: Governing youth mobility through vulnerability. *Journal of Ethnic and Migration Studies, 47*(5), 988–1005.

Helms, E. (2015, December 22). Men at the borders. Gender, victimhood, and war in Europe's refugee crisis. *Focaal Blog*.

Henley, J. (2020, July 9). LGBT asylum seekers' claims routinely rejected in Europe and UK. *The Guardian*. https://www.theguardian.com/uk-news/2020/jul/09/lgbt-asylum-seekers-routinely-see-claims-rejected-in-europe-and-uk

Herz, M. (2019). "Becoming" a possible threat: Masculinity, culture and questioning among unaccompanied young men in Sweden. *Identities, 26*(4), 431–449.

Hodal, K. (2020, February). 'A step away from hell': The young male refugees selling sex to survive. *The Guardian*. https://www.theguardian.com/global-development/2020/feb/21/a-step-away-from-hell-the-young-male-refugees-selling-sex-to-survive-berlin-tiergarten

Holloway, K., Stavropoulou, M., & Daigle, M. (2019). *Gender in displacement. The state of play*. Humanitarian Policy Group.

Honeyball, M. (2016). *Report on the situation of women refugees and asylum seekers in the EU (2015/2325 (INI))*. European Parliament.

Hudson, V. (2016, January 5). Europe's man problems: Migrants skew heavily male-and that's dangerous. *Politico*.

Hyndman, J. (2010). Introduction: The feminist politics of refugee migration. *Gender, Place and Culture, 17*(4), 453–459.

Hyndman, J., & Giles, W. (2017). *Refugees in extended exile. Living on the edge*. Routledge.

Indra, D. (1999). Not a "room of one's own": Engendering forced migration knowledge and practice. In D. Indra (Ed.), *Engendering forced migration: Theory and practice* (pp. 1–22). Berghahn Books.

Ingulfsen, I. (2016, January 5). Europe's main problems: Migrants skew heavily male -and that's dangerous. *Politico*.

Jansen, S., & Spijkerboer, T. (2011). *Fleeing homophobia: Asylum claims related to sexual orientation and gender identity in Europe*. COC Nederland, Vrije Universitei.

Johnson, H. L. (2011). Click to donate: Visual images, constructing victims and imagining the female refugee. *Third World Quarterly, 32*(6), 1015–1037.

Kačapor-Džihić, Z., & Oruč, N. (2012). *Social impact of emigration and rural-urban migration in central and Eastern Europe*. Final Country Report Bosnia and Herzegovina.

Kingsley, P. (2015, November 24). Canada's exclusion of single male refugees may exacerbate Syrian conflict. *The Guardian*.

Kofman, E. (2002). Contemporary European migrations, civic stratification and citizenship. *Political Geography, 21*(8), 1035–1054.

Kofman, E. (2019). Gendered mobilities and vulnerabilities: Refugee journeys to and in Europe. *Journal of Ethnic and Migration Studies, 45*(12), 2185–2199.

Kreis, R. (2017). #refugeesnotwelcome: Anti-refugee discourse on Twitter. *Discourse & Communication, 11*(5), 498–514.

Lewis, R., & Naples, N. (2014). Introduction: Queer migration, asylum, and displacement. *Sexualities, 17*(8), 911–918.

Luna, F. (2019). Identifying and evaluating layers of vulnerability – A way forward. *Developing World Bioethics, 19*, 86–95.

Macklin, A. (1999). Comparative analysis of the Canadian, US and Australian directives on gender persecution and refugee status. In D. Indra (Ed.), *Engendering forced migration: Theory and practice* (pp. 272–307). Berghahn Books.

Morris, L. (2002). *Managing migration*. Psychology Press.

Muftić, L., & Bouffard, K. (2008). Bosnian women and intimate partner violence differences in experiences and attitudes for refugee and nonrefugee women. *Feminist Criminology, 3*(3), 173–190.

Myrttinen, H, Khattab, L. and Maydaa, C. (2017). 'Trust no one, beware of everyone.' Vulnerabilities of LGBTI refugees in Lebanon. In J. Freedman et al. (eds) A gendered approach to the Syrian refugee crisis, pp. 61–76. : Routledge.

Oxfam. (2016). *Gender analysis. The situation of refugees and migrants in Greece*. Oxfam.

Ozcurumez, S., Akyuz, S., & Bradby, H. (2020). The conceptualization problem in research and responses to sexual and gender-based violence in forced migration. *Journal of Gender Studies*. https://doi.org/10.1080/09589236.2020.1730163

Parrs, A. (2018). The vulnerably refugee woman, from Damascus to Brussels. In C. Timmerman, M. L. Fonseca, L. v. Praag, & S. Pereira (Eds.), *Gender and migration: A gender-sensitive approach to migration dynamics* (pp. 217–242). University of Leuven Press.

Peroni, L., & Timmer, A. (2013). Vulnerable groups: The promise of an emerging concept in European Human Rights Convention law. *International Journal of Constitutional Law, 11*(4), 1056–1085.

Pisani, M., & Grech, S. (2015). Disability and forced migration: Critical intersectionalities. *Disability and the Global South, 2*(1), 421–441.

Pittaway, E., & Bartolomei, L. (2018). *From rhetoric to reality: Achieving gender equality for refugee women and girls*. Centre for International Governance Innovation.

Plambech, S. (2017). Sex, deportation and rescue: Economies of migration among Nigerian sex workers. *Feminist Economics, 23*(3), 134–159.

Pruitt, L., Berents, H., & Munro, G. (2018). Gender and age in the construction of male youth in the European migration 'crisis'. *Signs: Journal of Women in Culture and Society, 43*(3), 687–709.

Rettberg, J., & Gajjala, R. (2016). Terrorists or cowards: Negative portrayals of male Syrian refugees in social media. *Feminist Media Studies, 16*(1), 178–181.

Rigo, E. (2017). Re-gendering the border: Chronicles of women's resistance and unexpected alliances from the Mediterranean border. *ACME: An International Journal for Critical Geographies, 16*(1), 173–186.

Rohwerder, B. (2018). *Syrian refugee women, girls, and people with disabilities in Turkey*, K4D.

Scheibelhofer, P. (2017). "It won't work without ugly pictures": Images of othered masculinities and the legitimisation of restrictive refugee-politics in Austria. *NORMA, 12*(2), 96–111.

Scheibelhofer, P. (2019). Gender and intimate solidarity in refugee-sponsorships of unaccompanied young men. In M. Feischmidt, L. Pries, & C. Cantat (Eds.), *Refugee protection and civil society in Europe* (pp. 193–220). Palgrave Macmillan.

Sozer, H. (2020). Humanitarianism with a neo-liberal face: Vulnerability intervention as vulnerability redistribution. *Journal of Ethnic and Migration Studies, 46*(11), 2163–2180.

Squire, V., Dimitriadi, A., Perkowski, N., Pisani, M., Stevens, D., & Vaughan-Williams, N. (2017). *Crossing the Mediterranean Sea by boat: Mapping and documenting migratory journeys and experiences*. Final Project Report. www.warwick.ac.uk/crossingthemed

Turner, B. (2006). *Human rights and vulnerability*. Penn University Press.

Turner, L. (2017). Who will resettle single Syrian men? *Forced Migration Review, 54*, 29–31.

Turner, L. (2019a). Syrian refugee men as objects of humanitarian care. *International Feminist Journal of Politics, 21*(4), 595–616. https://doi.org/10.1080/14616742.2019.1641127

Turner, L. (2019b). The politics of labelling refugee men as "vulnerable". *Social Politics: International Studies in Gender, State and Society*, jxz033. https://doi.org/10.1093/sp/jxz033

UNHCR. (1991). *Guidelines on the protection of refugee women*. UN Doc. ES/SCP/67. Online: www.unhcr.org/3d4f915e4.html

UNHCR. (2002). *Guidelines on international protection: 'membership of a particular social group' within the context of Article 1A(2) of the 1951 Convention and/or its 1967 Protocol relating to the Status of Refugees.*

UNHCR. (2008). *UNHCR guidance note on refugee claims relating to sexual orientation and gender identity*

UNHCR. (2012). *Guidelines on International Protection No. 9: Claims to Refugee Status 1951 convention and/or its 1967 Protocol relating to the status of refugees.*

UNHCR. (2013). *Response to Vulnerability in Asylum – Project Report*. http://www.refworld.org/docid/56c444004.htm. Accessed December.

UNHCR. (2014). *Introducing the vulnerability assessment framework.*

UNHCR, UNFPA and Women's Refugee Commission. (2016). *Initial assessment report: Protection risks for women and girls in the European refugee and migration crisis.* Greece and the Former Yugoslav Republic of Macedonia.

UNHCR. (2016a). *Questionnaire findings for Afghans in Greece.* May.

UNHCR. (2016b). *Questionnaire findings for Syrians in Greece,* May.

UNICEF. (2020). *Latest statistics and graphics on refugee and migrant children Latest information on children arriving in Europe.* https://www.unicef.org/eca/emergencies/latest-statistics-and-graphics-refugee-and-migrant-children. Accessed 11 Aug.

Uteng, T., & Cresswell, T. (Eds.). (2008). *Gendered Mobilities.* Ashgate.

Williams, L., Coskun, E., & Kaska, S. (Eds.). (2020). *Women, migration and asylum in Turkey. Developing gender-sensitivity in migration research, policy and practice.* Palgrave Macmillan.

Women's Refugee Commission. (2016a). *No safety for refugee women on the European route.* Report from the Balkans. Women's Refugee Commission.

Women's Refugee Commission. (2016b). *Falling through the cracks: Refugee women and girls in Germany and Sweden.* Women's Refugee Commission.

Yuval-Davis, N., Wemyss, G., & Cassidy, K. (2019). *Bordering.* Polity.

Zlotnik, H. (2003). *The global dimensions of female migration.* March: Migration Information Source.

Chapter 6
Engendering Integration and Inclusion

In this chapter we turn to issues of how migrants participate in society and especially their gendered aspects. Why is gender important in this regard? It is a consideration that is usually absent from both theoretical and policy discussions of what is commonly termed *integration* or the basis on which migrants are incorporated into a society, a term widely used across different societies but with different meanings (Rytter, 2019). Whilst integration policies might seem to be neutral, they may in effect target women and men differently and have different outcomes for them. Such policies may also apply primarily to certain categories of migrants, although the categories and nationalities change over time. As we shall see, concerns over what constitutes problematic integration vary, such as: lack of knowledge of the language of the country, non-participation in the labour market and traditional cultural and social practices transferred from societies of origin. These have generated demands to impose integration measures and contracts as conditionalities of immigration and, if applicable, to the different stages in the pathway to citizenship.

In the first section we examine an increasingly critical debate on the notion of integration. This debate has probably been more visible in academic writing than in policy interventions, where discourses of securitisation, targeting of Muslim populations, their unwillingness to 'integrate' (Kontos, 2014) and the retention of transnational ties and practices, prevail. The relationship between academic discourses and policies is a difficult one in which the critical edge of academic studies may be lost (Rytter, 2019). There has also been, with a few exceptions (Anthias & Pajnik, 2014; Korteweg & Triadafilopoulos, 2013), little reflection on gendered or intersectional aspects. For Schinkel (2018: 4) class and race have been purified from integration and, we would add, so too have gender and sexuality.

In the second section, we show how integration discourses are gendered in the way they represent and target migrant women and men (Anthias & Pajnik, 2014; Kofman et al., 2015). Those entering through family migration routes (Bonjour & Kraler, 2015; OECD, 2017), whom as we have seen in Chap. 4 are in the majority women, were the first to be subjected to integration measures. In terms of integration, women are supposedly reluctant to 'integrate' and 'become one of us', while men's

© The Author(s) 2022
A. Christou, E. Kofman, *Gender and Migration*, IMISCOE Research Series,
https://doi.org/10.1007/978-3-030-91971-9_6

patriarchal culture, especially if Muslim, holds women back and is dangerous for contemporary values of gender and sexual equality (see also Chap. 5 in relation to recent discourses on young refugee men). In effect, in gendering, racializing and classing certain categories, it is the 'migrant with poor prospects' (Bonjour & Duyvendak, 2018) who must be forced to integrate. In earlier years of proposals to implement integration measures in particular, a gendered argumentation was evident in a number of states (Kofman et al., 2015; Korteweg, 2017). And as a number of scholars have critically commented, the analysis of what is to be done in relation to integration strips migrants of any heterogeneity or probes the relationship between the different aspects of their subjectivity i.e. intersectionality (Korteweg, 2017) or problems the receiving society puts in the way of their insertion (Korteweg & Triadafilopoulos, 2013). Women's bodies have become the battleground, for example through the targeting of garments worn by Muslims as inimical to Western liberal values and a threat to security. At the same time, the skilled are depicted as unproblematic in terms of their integration and hence not requiring any support to settle (Weinar & Klekowski von Koppenfels, 2020: 3–4).

6.1 Immigration and Integration: Insights and Debates

Migration scholarship has applied a number of terms to denote processes by which immigrants participate or not in the new societies they move to, the degree and context of such participation, the impact on their identities and their offspring's identities and so on. The terminology includes such concepts as 'integration', 'incorporation', 'assimilation', 'acculturation', as well as the evolving phenomena transforming societies termed 'multiculturalism', 'cosmopolitanism' and recently, 'superdiversity'. Theorists (Brubaker, 2001; Favell, 2003; Schinkel, 2018) have argued about the validity, usefulness and limitations of these concepts and have exemplified their usage in particular historical and geographical zones (Alba & Foner, 2015).

Indeed, integration has been reconceptualised 'as a two-way process' and as such conceived as 'primarily a matter of social standing, and not legal or socioeconomic status' (Klarenbeek, 2019: 2). In this sense the evaluative standard is viewing society with no social boundaries demarcating participation between those legitimised as members and thus societal insiders and those non-legitimised members and hence outsiders. In another sense, the grappling between the need for migrants to maintain their ethnic differences and cultural identities and the functionality of becoming 'absorbed' in their new place of residence points to the ability to interact on a normative equal level with those of the majority population and established as residents. In other words, according to Castañeda (2018: 2–3) 'integration means upward social mobility, no residential segregation, intermarriage, and the potential for equal participation in politics and public activities. …Unlike assimilation, integration does not imply losing the culture of the country of origin but actually being able to sustain it while also adapting to a new city'. Castañeda finds the term integration to be similar to the concept of *incorporation* in pointing to the inclusion of excluded groups into the political and social structures of a given city and thus important in understanding feelings of belonging for migrants to the cities where they are residing. From a more intersectional perspective, the notion of integration for some scholars produces gendered and racialised non-belonging (Korteweg, 2017).

Furthermore, others have underscored that the concept 'further risks concealing and perpetuating power dynamics and (colonial) hierarchies' (Meissner & Heil, 2020: 1). Such hierarchies contribute vastly to social differences and the relational practices, power asymmetries and everyday life materialities that create group dynamics and reconfigure social relations. As a result, some researchers find more analytical purchase and conceptual utility in the concept of 'disintegration' (ibid) as a provocation to moving away from integration processes and the examination of group or individual performance in a given society.

However, policy aspects of integration impact on immigrant labour, families and refugees. Recent literature (Birkvad, 2019) delineates the significance of citizenship in a 'Western' country of residence as naturalisation has important meanings for: mobility which in turn facilitates transnational connections; denoting legal stability offering a sense of security for those in precarious situations and liminality of legality; designating a formality of recognition of equality and belonging. Such meanings placing importance on citizenship in achieving migrant access to mobility, stability and recognition combine both strategic and more instrumental conceptualisations aligned to more emotive and symbolic meanings around civic status.

Some of the conceptualisations we discussed in Chap. 2 (affective, performed/ embodied, intersubjective and spatial) are key dimensions in how the key concept of 'lived citizenship' has been applied in the last two decades since its emergence (Kallio et al., 2020). Lived citizenship has been conceived as a 'locus of political agency in participatory policy' (Kallio et al., 2015) while acknowledging the spatial complexities in transnational mobilities (Wood, 2013) and the emotional geographies of citizenship participation. In this direction, an intersectional lens can be applied to contesting integration (Anthias & Pajnik, 2014) as an interconnected analytical venture with larger social inequalities and hierarchies within transnational experiences and perceptions of migrants as active social agents in contemporary host societies. This inevitably links integration discourses with migrant experiences and negotiations in their everyday life practices and instances of marginalisation. This approach points to the way gender hierarchies are intertwined with gendered social relations of power and renders integration a highly normative and problematic concept (Anthias, 2013; Anthias et al., 2013).

Approaches to the study of migrant integration are multiple and range from focusing on social, political, cultural or economic integration and even expand to literatures in the second and subsequent migrant generations. In this chapter, the gendered focus on integration and citizenship is on contemporary issues regarding policy matters of immigrant generations primarily in the European landscape but with a few global examples supplementing the debates. From a policy objective, the foci are primarily on interventions and practices that make integration successful or problematic. While meanings of successful and unsuccessful integration might diverge, the emergence of an increasingly hostile environment, everyday bordering and efforts towards activism in combating such phenomena suggest that by and large policy approaches to integration require re-thinking (Yuval-Davis et al., 2018).

While migration has always been a divisive political issue as it concerns issues of sovereignty and identity, the increasing ethno-cultural diversity of receiving societies requires governments to find ways to respond to such changes. Subsequent immigrant generations have grown up protesting exclusion, racism and discrimination in the societies where they claim their right to equal opportunities. Political tension and conflictual clashes have evolved in the form of riots such as those in the UK and France over the last two decades. At the same time, there are politicians and media outlets who shift the blame to the migrants themselves by underscoring their failure to integrate in privileging their distinct cultures and religion, thus jeopardising social cohesion and becoming a threat to national security (de Haas et al., 2019; Korteweg & Triadafilopoulos, 2013).

More reflexive approaches to the study of integration have advocated a shift outside the normalisation discourse in order to disentangle research from the migration apparatus. This has led to methodological strategies to de-naturalise and de-ethnicise integration (Amelina & Faist; 2012; Levitt, 2012) and even to the more ambitious approach to 'de-migranticize' integration studies to be more reflexive in this regard by distinguishing analytical and commonsense categories, by aligning social theory to research on migration in order to remove the 'migration container', and by challenging the object of study from the entire migrant population to segments of the overall population (Dahinden, 2016). It appears that Dahinden (pp. 2219–2220) does not argue either for 'more' or for 'fewer' migration and integration studies, but for different ones which would reconcile these contradictions through following a triple strategy, of firstly, centring migration studies within social theory; secondly, moving out of 'migration containers' to alignment with and within social science; thirdly, 'migranticizing' general social research by embedding ethnic and migration studies into disciplinary university curricula. This, Dahinden calls, a plea to establish a 'post-migration' social science that embraces migration and integration transversally within social science research and theory.

While these elements are sound endeavours to a more inclusive alignment of migrant integration approaches and social theory, social science research and academic curricula, there appears to be an oversight of a more gendered *and* intersectional engagement which would address some of the extractive and power laden entanglements of who is 'integratable', who is not, who is more, who is less, why and when some groups are allowed to integrate and others are continuously marginalised. A gendered and intersectional awareness of these issues also has implications about the ways we teach our curricula, the research that informs and drives the pedagogies we produce, what gets to be on those curricula and what is excluded, who teaches what, what kind of research is conducted, what is published and receives funding to co-produce gendered and intersectional knowledge on these themes. Thus, are we yet entering additional spaces of neo-colonialities of epistemic communities if we don't consciously engage with a gendered and intersectional approach within and beyond integration. But this does not mean that just as colonialism expertly abused nations and people for raw materials, diverse epistemologies should also become extractive mines for the accumulation of knowledge and the development of theory in the global North. The intellectual and epistemic reparations we envision are those dismantling existing epistemic hierarchies to prevent the reproduction of similar damaging dynamics. We need to think about what knowledge has been denied or silenced and how we give voice, not in a paternalistic context, but a critical, uncomfortable, and meaningful way to think differently about dominant knowledge. Perhaps future research is what is needed to claim new spaces of knowledge production which are not superficially dismissive of integration, but instead, seek to situate its limitation and harness its public policy potential.

In terms of the public policy and activist potential for migrant groups, more novel interpretive frameworks use the lens of 'communities on the move' (CoM) and focus on the idea of such cultural communities capitalising on shared values and network ties which produce knowledge and opportunities to facilitate integration (Parrilli et al., 2019). In a sense this is the next phase of the 'social capital' migration theories approach which underscores the capacities of mobile communities to shape migrant well-being and innovation through knowledge production aiding integration. In essence, this approach focuses on the bonding and bridging of migrant groups as a complementary performance to embed them into the regional context.

Other studies show that national narratives of negative expatriate nationalism might even prevent integration into the host society. The study of Isaakyan & Triandafyllidou (2014) on Anglophone marriage-migrants in Southern Europe focusing on expat nationalism and integration dynamics has revealed that their own understanding of themselves as 'long-term visitors' evaluating an 'inferior' culture of the country they are 'visiting', coupled with their Anglophone culture of origin, leads to them coining the term of 'broken integration'. This term is a useful heuristic device to explain the challenges of adaptation, interaction and ultimately integration within the host country culture. This is a particularly intriguing insight into the reverse kinds of power dynamics that can stimulate nationalisms that obliterate pathways to integration when migrants perceive themselves in a cultural hierarchy above the host society which they consider inherently inferior. Ultimately, this is also 'resistance agency' on the part of migrants who perceive 'broken integration' as their sole option in accepting their settlement as inherently problematic because of their superiority and not that of the majority residents.

Those studies continuing and contributing to the strand of literature that bridges social, political and acculturation psychological theories on immigrant integration aim to disentangle how the objective of citizenship and the subjective of perceived social status and belonging correspond in different societal contexts to socio-political integration. These studies examine integration as a mutual intergroup process between immigrants and the receiving society and utilise a person-oriented approach, such as the study by Renvik et al. (2020) with their survey exploring the patterns of socio-political integration among Russian-speaking minority group members in three neighbouring countries in the Baltic area: Estonia, Finland, and Norway. While three profiles were obtained in all countries (critical integration, separation and assimilation) that of 'critical integration' was the most common and when examined in relation to the citizenship and integration context.

Finally, extending the concept of 'critical integration' to delving into more critical explorations of the utility of the concept of integration, in a recent themed issue of the journal *Comparative Migration Studies* (2019) on debating integration as a central, yet increasingly contested, concept, Willem Schinkel's (2019) article triggered a flurry of responses and then a rejoinder by the author concluding that migration studies should be seen as an 'imposition'. Schinkel's initial paper entitled, 'Against "immigrant integration": For an end to neocolonial knowledge production' (2018), based on his book *Imagined Societies. A Critique of Immigrant Integration in Western Europe* (2017), and intended as a provocation piece, outlined three core arguments: firstly, immigrant integration research has been lacking robust conceptual grounding and in particular sociologically rigorous notions of core concepts such as 'society'; secondly, the very process of monitoring immigrant integration is a form of neo-colonial knowledge production deeply entangled with contemporary workings of power; and, thirdly, makes a bold proposition for a more critical social science perspective, one that bypasses notions of 'immigrant integration' and 'society' in perhaps embracing more of a sociological imagination. This approach would focus on the actuality of migrants crossing social ecologies without the deterministic aspects of policy categories or commonsense explanatory frameworks.

Penninx (2019) in his response signals three building blocks as alternative solutions to the problems that Schinkel advances with the concept of integration. The first proposed is that research uses the broadest heuristic definition of 'processes of integration' as analytical concept to study a threefold approach (individual, collective and institutional) of a threefold dimensional context (legal, socio-cultural/economic and religious) of interaction between migrants and receiving society. The second proposition is to distinguish the study of integration policies as fundamentally distinct from the processes of integration and hence key questions should be framed to reflect perceived causes and solutions. The third component of an alternative practice is rather a sensitive one alluding to pressures in 'safeguarding scientific independence against mounting pressures on programming and content of research – often through funding of research' where the challenge is seen as the strong politicisation of the topics of migration and integration, as can be seen in the ways in which the need for integration of selected migrant categories are framed (Kofman et al., 2015; Korteweg, 2017).

Adrian Favell (2019) proposes a series of 12 propositions to rethink the utility of 'integration' as a concept, given that it is deeply embedded in methodological nationalism and by extension produces colonial, nation-state centred visions of societies while sustaining inequalities and orders of social power hierarchies (Anthias & Pajnik, 2014; Korteweg & Triadafilopoulos, 2013). In the next section we explore in greater depth how integration policies have been framed in the past two decades (Eggbø & Brekke, 2019), highlighting the key role of gender and sexuality in othering non-Western populations and regulating their family practices.

Whilst for national populations not requiring integration (Schinkel, 2018), family life has become a matter of individual responsibility, for families with a migrant background it requires a tutorial state (van Walsum, 2012: 6).

6.2 Integration Policies, Gendered Interventions and Outcomes

National policies on migrant integration in the EU have emerged within regimes and debates shaped by a 'Euro-crisis' that sees fundamental disagreement about migration issues at large and certain aspects of the debate constructed in discourses of moral panics, deviance, crime and securitisation (Trimikliniotis, 2014). This is also a reflection of how historical contexts have shaped governance practices in different EU countries and by extension integration agendas reflect distinct agendas and institutions in turn shaped by politics, history, and culture (Hernes, 2018). As such, Nicos Trimikliniotis (ibid) suggests we need to map integration agendas as we map contestations about the meaning and priorities of integration while locating the debates in the neoliberal transformations taking place.

As a consequence of a neoliberal conceptualisation of integration and the implicit accusation of migrants as 'unwilling to integrate' into their host societies this has led to a normative framing of action in the form of the 'integration contract'. The latter has been implemented in a number of European countries such as the Netherlands since 2002, Austria since 2003, France and Denmark since 2006, Luxembourg and Germany since 2011. This constitutes a form of an explicit agreement between states and migrants setting out what measures and support the state offers and what needs, responsibilities and expectations the migrants have in order to integrate. The compelling of compliance with the core ('westernised/democratic') values of the respective society are summarised as the basics of gender relations, diversity, equality and freedom of speech. In an assimilationist understanding of cultural negotiations of values and expectations, often women's rights and compliance with particular understandings of gender equality are instrumentalised for a synthetic subordination (Kontos, 2014; Kostakopoulou, 2014).

While gender relations and inequality had largely passed without comment in discussions of the crisis of multiculturalism in academia or in the media (Phillips & Saharso, 2008), migrant women moved from the invisible periphery to the all too visible core (Prins & Saharso, 2008) in discourses around the need to impose integration measures for migrants.

It can be seen most clearly in the conflicting positions adopted around the veiling of Muslim women. Opposition to the wearing of the headscarf by Muslim women emerged as a major political issue, especially in France where it was seen as undermining secular values, in the 1990s. Some feminist voices have highlighted the continuation of colonial practices of controlling colonised bodies (Bassel, 2021) evident in this intervention in which the colonial trope of unveiling women as liberation prevails (Scott, 2007), and the role of liberal feminism in refusing to acknowledge the voices and demands of Muslim women (Korteweg & Yurdakul, 2020). The oppressed Muslim woman is pitted against the opposite of the free, gender-equal citizen in what Dahinden and Manser-Egli (2021) have called "gendernativism", a "gendered and racialized form of xenophobia that constructs the 'Other' as the opposite of the free, gender-equal, 'real', 'authentic', 'rooted' citizen".

The arguments for and against the banning of different forms of veiling (headscarf, niqab, burqa), its contestation, the role of the state, the place of religion in the public sphere, and the recognition of diverse practices are too complex to address in this chapter. In the box below on the comparison of headscarf debates we briefly outline the legal and political tussles in states in which the issue has been politicised. Courts have more often adopted rights-based perspectives while the political realm has focussed more on the issue in terms of integration, national unity, expression of political Islam and security (Joppke, 2009).

Box: Comparison of Headscarf Interventions

Only a few states in the European Union (Croatia, Cyprus, Greece, Poland, Portugal, and Romania) have not implemented or debated proposals to ban the headscarf either in public space in general or more commonly in specific sites, such as schools, or among particular occupations such as teachers or civil servants (Weaver, 2018). France is the most restrictive and after 15 years of contestation since 1989, over what was known as *l'affaire du foulard'*, Parliament passed overwhelmingly in 2004 a law banning the wearing or display of ostensible religious symbols in public schools, seen as the crucible of Republican values. In Germany it was teachers as civil servants who were targeted by many regions. In 2003 in response to a teacher (Fereshta Ludin), who had been prevented in 1998 from working wearing a headscarf, the Constitutional Court ruled in her favour but passed the matter to the political level stating that those Länder seeking to prohibit the wearing of the veil had to pass legislation. However in 2015 the Federal Constitutional Court issued a clarification stating that if 'there is no concrete danger to disrupting the peace at school or the state's duty of neutrality, headscarves are permissible for women in the teaching profession' (Chang, 2021).

In addition the firing of women for wearing the headscarf in workplaces has also been challenged. Initially the European Court of Justice concluded in a ruling (2017) of cases brought by Muslim women in Belgium and France that the employer could ban an employee from wearing visible religious symbols as long as they had a policy in place (Weaver, 2018). However a more recent judgement by the European Court of Human Rights concluded that a Belgium court had no right to insist that women appear uncovered in court, which was

(continued)

the first time this court has ruled in favour of the right of Muslim women to be veiled (Cox, 2018).

In terms of face veiling, France and Belgium banned in 2011 the niqab and the burqa in public which has been followed in other countries such as the Netherlands in 2015 (only on public transport and public areas but not on the street), Austria (2017) and Denmark (2018). Actual bans or proposals to ban the wearing of the full veil have happened in almost all European countries (Open Society, 2018), the latest being supported in the Swiss referendum in March 2021.

In terms of interventions relating to immigration and integration, one can discern several, often overlapping, discourses from the beginning of the century.

1. **Gender inequality** in relation to work where migrant women have low rates of participation in the labour market which has generally been the core element of equality. In Scandinavia, the emphasis on emancipation and independence is to be achieved through the labour market (Eggbø, 2010). A satisfying family life would be achieved by women working and independently earning their own income (Bech et al., 2017) working was an integral part of being a good citizen. In other countries, such as the UK, the desirability of labour market independence might be more about reducing reliance on public services. As has been pointed out (Korteweg & Triadafilopoulos, 2013), there was little consideration of discrimination faced by migrant women in the labour market or the actual level of participation of different groups of migrants, some of whom had higher levels of participation than native women. Instead the focus has been on Muslim women from Morocco and Turkey in Germany and the Netherlands, and from Bangladesh and Pakistan in the UK.

2. The **failure of integration** due to socio-economic marginalisation and the formation of an ethnic underclass arising in part from poorly educated spouses who as mothers do not have the requisite skills to educate their children to succeed in society and hence contribute to the continuing reproduction of socio-economic inequalities (Joppke, 2007). This emerged quite clearly in the Netherlands. The motto 'if you educate a woman, you educate a family' was used in the Dutch PaVEM Commission (Prins & Saharso, 2008) where women were seen as the reproducers of the next generation and thus required a better start (Kirk, 2010). Dutch parliamentary debates mentioned that the marginalisation of specific population groups could be passed from generation to generation, hence the need to ensure that women have a better starting position in the Netherlands (cited in Bonjour & de Hart, 2013). Though most explicitly stated in the Netherlands, this theme of the social reproduction of family members (love, marriage, parenthood, fertility, adult dependency) as future citizens would implicitly underpin a series of future regulatory controls over intimate and family relationships seeking to steer migrants' belonging to the nation (Bonizzoni, 2018.)

3. **Family practices** incompatible with liberal societies and the formation of couples within transnational marriages (see Chap. 4). Western 'liberal' and open societies had to be protected from patriarchal and traditional gender roles where the body of the female Muslim migrant served to demarcate the boundary between the civilised Westerner and the uncivilised illiberal outsider (Kirk, 2010; Razack, 2004). Gay emancipation was also mobilised to frame Muslims as non-modern subjects (Mepschen et al., 2010). These illiberal practices included forced marriages, honour killings and transnational marriages with cousins, a particular concern in Denmark.

There was widespread agreement that the problem lay in the laxity of family reunification policies (Schmidt, 2011) and high levels of transnational marriages, hence family migration became the terrain for the control of cultural differences beginning with admission but extending to further stages of permanent residence and citizenship. Hence, dealing with forced marriages generated demands for language proficiency prior to entry. Thus Ann Cryer, at the time a Labour MP for a constituency with a large Muslim Asian population, made a direct connection between arranged marriages, difficulties in learning English and the success of different ethnic communities in the UK and thus called for English tests (Kofman et al., 2015). In Germany too it was argued that those caught up in forced marriages were prevented from leading an independent life because of poor language proficiency (Yurdakul & Korteweg, 2013) and resisting parental authority and other family pressures (Lechner, 2011). Initially men were not envisaged as being affected by forced marriage though in the UK gay men were subsequently included (Samad, 2010). In Denmark, politicians conceived of forced marriage as primitive and 'un-Danish' with no place in the country (Schmidt, 2011: 362–3).

Whilst certain attitudes were shared, policy responses differed. An understanding of options pursued would need to take into account the strategizing deployed by politicians, NGOs and individual feminists and the interplay of political forces in a particular context. As Hagelund (2020) suggests, migration studies require more focused attention to discursive processes of policy construction in understanding how integration policies vary, not just in sense-making, but also in the implications for policy choices in crafting policy agendas.

From about 2005 to 2006, three major policy initiatives ensued in a number of countries in North western Europe. These were implementation of language and, in some cases knowledge of society pre-departure tests; raising the age of marriage; and the imposition of a minimum income for the sponsor to be able to bring in a spouse or children.

Language tests were adopted in a number of countries (Austria, Denmark, Germany, the Netherlands, the UK) (Goodman, 2011). In the Netherlands an analysis of the effects of the application of the language test of an A1 knowledge in Dutch as from 2006 showed that it had upgraded the human capital of the entrants but this may have been as much the effect of the selection of a spouse who was capable of passing the test (Scholten et al., 2012). Their effect was to change the socio-economic composition of family migrants and/or reduce the numbers entering through family migration.

Though **forced marriage** had played a part in the rationale for language tests, the predominant argument came to be the improved chances for integration. It was in relation to an increased age in marriage where the argument of prevention of forced marriage dominated. The targeted political subject was the Muslim woman but in order to avoid accusations of discrimination no one was exempt, including partners with non-migrant backgrounds. Only in the UK was the lifting of the age of partners in overseas marriages from 18 to 21 years in 2008 successfully challenged in the courts and lifted in 2012. The judgement found that imposing a blanket rule in order to deal with about 4% potentially of forced marriages was unjustified and disproportionate. Age of marriage was raised first in Denmark to 24 years in 2003 where it was strongly felt that intervention in the private sphere was seen as appropriate to ensure conformity to social norms (Fog Olwig, 2011). Furthermore, an attachment condition stating that one had to have more links with Denmark than with any other country was stipulated. In Norway, on the other hand, although forced marriages were hotly debated in the Immigration Act Commission of 2004, policy attempts to regulate forced marriages among Pakistani and Turkish populations did not take the route of raising the age of marriage for migrants, which was scrapped in 2007, but through imposing a minimum income rule (Eggbø, 2010; Staver, 2015).

Minimum income regulations represent the drift towards economic imperatives and links labour market participation to family migration (Kofman et al., 2015; Staver, 2015; Sirriyeh, 2015), in this case by the sponsor. As such, it reflects the transposition of economic criteria normally demanded of skilled migrants to family migration based on normative principles. In an increasingly neo-liberal immigration policy, not only migrants but those sponsoring migrants, are expected to demonstrate that they are able to be autonomous and responsible for themselves (Schinkel & van Houdt, 2010) and not be a burden on welfare services. In countries that imposed the highest income criteria – the Netherlands in 2004, Norway in 2010 and the UK in 2012 – the latter two demand high income levels in which only the individual, and not the family as a whole, can provide the necessary resources, that is the individual must demonstrate that he/she is responsible for themselves. The outcome is to produce a class and gendered and discriminatory impact. Other countries have since then also adopted minimum income requirements (European Migration Network, 2017).

Let us take the case of the UK to probe the gendered and intersectional possibilities in the right to family life. The objectives of the policy were: 'ensure that migrants are supported at a reasonable level that ensures they do not become a burden on the taxpayer and allow sufficient participation in everyday life to facilitate integration' (Home Office, 2011a). The premise is that low income British citizens as sponsors would lead to difficulties of integration. It's clear that the unspoken assumption is that sponsors would be either settled migrants or descendants of migrants, and probably of South Asian origin. When the original level of minimum income of £18,600 was introduced in July 2012, it was set at 140% of the minimum wage, at which level it was calculated that the couple would not qualify for any income-related benefits. The Migration Observatory estimated at the time that 47% of British citizens in employment would not qualify as sponsors but that women, certain minority ethnic groups, especially Bangladeshis and Pakistanis, young people between 20 and 30 years and those living outside of London and the South East would be disproportionately affected.

Analysis of Labour Force Survey data for 2012–2017 confirms that women are the most affected followed by ethnic minorities overall (Sumption & Vargas-Silva, 2019). In general the gender pay gap, concentration in low paid and greater propensity to work part-time, and their caring responsibilities, mean that women's annual incomes are lower than men. However when we combine gender and ethnicity using data from the January to March 2017 UK Labour Force survey a much more complex picture emerges. The median salary of women employed full-time is about 88% of the UK median income. It is slightly lower for white women, much higher for Chinese, Indian and the category of other ethnic, but substantially lower for Bangladeshi and Pakistani women. Including part-time work as well pushes the median below £18,600 for everyone except the group Mixed Ethnic, itself a very heterogeneous category. Thus family migration is only a possibility for full-time workers.

An interesting dimension of this requirement was that as a result of the increasing precarity, even middle class highly educated citizens could also be caught in its tentacles. Global mobility by working holiday makers, students and workers has led to an expansion of intimate relations and partnerships (Wagner, 2015) complicating stereotypes of transnational marriages. While 10 nationalities made up almost 50% of marriage migrants in the UK, a very large number of diverse nationalities comprised the rest (Home Office, 2011b). However, those with some flexibility and cultural capital turned to remedies available in international law, for example, exercising their free movement rights to go to another European Union country which allows them to move and reside freely with their spouse and children and even parents without any income requirement. A small scale study of 20 couples in the UK, who fell short of the stable minimum income, highlighted their high cultural capital and flexibility in relation to types of work and age of children (Wray et al., 2021). We do not know how common this strategy has been in the European Union, although as a result of Denmark's stringent attachment rule, it is estimated there are 2000–3000 Danes unable to reunify who have moved to neighbouring Sweden, especially those living in Copenhagen.

Though an attitude shared with other countries, Denmark probably exemplifies more than any other a fixation with transnational marriages by minority ethnic groups (van Kerckem et al., 2013) arising from the stringent attachment rule (Bissenbakker, 2019). From 2000 to 2018, the Danish attachment requirement stated that family reunification in Denmark could only be granted if the spouses' combined attachment to Denmark was stronger than the spouses' combined attachment to any other country (Ministry of Integration, 2002 [L152] §9, part 7). It had followed from an increase in the marriage age to 24 years for both partners but had effectively been raised to 28 and then dropped to 26 years as the length of the attachment period. Following a court ruling of the European Convention of Human Rights in 2016

(European Commission, 2019) finding the requirement to be discriminatory against some Danes, both major parties agreed to replace it with an integration requirement. However this makes the conditions even harsher with stipulations of language requirement for both sponsor and spouse together with three out of five other requirements. A draconian extension has been that the housing criteria has been expanded to cover not living in a ghetto defined as an area of non-profit social housing, in which 'the proportion of immigrants and descendants from non-western countries exceeds 50 percent' and in which, 'compared with the country as a whole, residents in the designated areas will generally have a significantly lower educational level, a weaker connection to the labor market and the educational system, a lower income, or have committed more criminal offenses' (Ministry of Foreigners and Integration 2018 [L231], 23). One wonders how long this extraordinary form of discrimination will survive without a legal challenge.

Today the integrationist imperative applies to other categories of the population, including skilled labour migrants, and refugees who are increasingly exhorted and compelled to integrate, particularly through participation in the labour market (Rytter, 2019; Schultz, 2020). The conditionalities of entry have also been extended to the pathway to citizenship, which are increasingly accompanied by criteria and requiring resources en route. Thus the path to citizenship is less than smooth with numerous hurdles of language, economic resources and dependency through the increasingly lengthy probationary period, potentially locking spouses into harmful relationships and subjected to gender violence in order to remain in the country (Briddick, 2020). The Council of Europe *Istanbul Convention preventing and combating violence against women and domestic violence* (adopted in 2011) covers asylum-seeking, refugee and migrant women. Chap. 7 Article 59 asks states to do all in their power to give partners who are dependent on their spouse for a residence permit to be given an autonomous residence permit in case of dissolution of the marriage due to violence irrespective of the duration of the relationship. However the Convention has not been signed by a number of countries, especially those in Eastern Europe and the UK. Furthermore, Poland notified in July 2020 its intention to withdraw which Turkey did officially on 22 March 2021.

6.3 Beyond 'Integration'? Activism and Inclusion

Overall, the concept of integration has been problematic with limited heuristic value when it shifts away from the functionalistic underpinnings that most policy-driven approaches offer. Reframing the concept in the direction of more democratising discourses which are transnational and intersectional can uncover some of its paradoxes away from its purely normative considerations (Anthias et al., 2013). To ameliorate its conceptual framing from an instrumentalising apparatus of domination over migrants and revitalise its meaning-making to one of social inclusion requires the consideration of critical citizenship studies combined with activism practice.

Migrant voices count, migrant experiences count and their agency is how these are actualised in activism. The mobilisation of migrant agency happens in both organised movements but also everyday acts of resistance and solidarity. These can be digital but also embodied. Karayianni and Christou (2020: 11) talk about 'new geographies of empowerment' in the digital era during which misogyny, sexism and gendered violence continue to explode and perceive feminisms of resistance in relation to social media as a renewed opportunity for activism. Embodied activisms are particularly pronounced with direct public interventions as in the case of the Athenian context where the socio-political forms of migrant squats and the socio-spatial interactions they foster and generate, represent not just sites, but also embodied practices for contesting citizenship (Raimondi, 2019). Raimondi (2019: 559) explores such migrant acts of resistance and activism by looking at them from a particular angle that draws on the 'gaze of autonomy' in also reinventing migrants and non-migrant activists in urban spaces.

As political struggles, these are historical strategies to gain access to urban space as a 'right to the city' (Lefebvre, 1996; Harvey, 2008), including gendered rights to participate and reclaim the city for migrant and minority women (Vacchelli & Kofman, 2018) whose claims to rights to the city and assertion of everyday citizenship take a number of forms. Muslim women, particularly those who are most visibly from the banlieue (suburbs) in the Paris Region, have experienced discrimination and harassment and a restricted "right to the city", particularly in affluent or middle-class areas, which has forced them to modify their use of transportation and shopping (Hancock, 2015; Hancock & Mobillon, 2019). However on International Women's Day eighth March 2015, some feminist groups staked a claim to central areas of Paris in marching in an alternative demonstration comprising women wearing a veil, lesbian, bisexual and queer persons and sex workers in opposition to the official march in which they were often placed at the end (Hancock et al., 2018). Another example of claiming urban rights are those engaged in heightened mobility, such as circular live-in-care workers, whose numbers have increased substantially in many European countries, have also fought through the courts and through collaboration with researchers and unions to overcome the lack of rights stemming from their in-betweenness and lack of protection for workers in household. In the example from Basel, Switzerland, they successfully gained recognition and financial compensation for the work they undertake. Beyond the successful outcome, Chau et al. (2018) note the way in which the gendering of the right to the city draws attention to the bridging of the divide between public space and private households beyond the more typical focus on formal citizenship and the public sphere.

Domestic work (see Chap. 3) has been increasingly politicised and an issue bringing together the global, national and the local in activism seeking to improve conditions of work and social protection (Cherubini et al., 2018; Mulally, 2015; Schwenken, 2017). Global networking (International Domestic Workers Network in 2006 subsequently Federation in 2012) together with collaboration with trade unions and institutions of global governance, namely the ILO led eventually to the passing of Convention 189 on domestic work in 2011. Though global in reach, it has only been ratified by 31 countries, largely in Central and South America

(17) and Europe (8) as of March 2021. However ratification has not necessarily brought mobilisations and enactment of rights. Cherubini et al. (2018) note that the Convention has tended to generate mobilisations and extended rights where it is embedded in prior local struggles and political projects and involves national workers and internal migrants from rural areas, as in the case of Colombia. Where it largely concerns international migrants, as in the case of Italy, which ratified the Convention in 2013, it has been treated as a bureaucratic matter with advocacy organisations not particularly visible.

In France, on the other hand, which has not ratified the Convention, the sector is much more regulated due to a national collective convention for salaried workers employed in the private sector since 1982 and updated in 1999 (Lepetitcorps, 2018). Furthermore undocumented workers are also recognised as having employment rights unlike in the UK and Ireland (Murphy, 2015). Though traditional trade unions in France have tended to treat activist groups in this sector as ethnic ones despite the fact they have moved beyond an initial network of those belonging to a single nationality or regional focus to one that embraces a wider spectrum of migrants. Lepetitcorps' study drawing on the experiences of two activists in this sector, notes their previous work experience in their countries of origin and their diversity of class backgrounds. Their engagement in political mobilisation stemmed from different trajectories. For the person from Mauritius it was to regularise her status and those of other domestic workers, for the middle class woman from the Ivory Coast it was to organise in a specific sub-sector of child minding, which was becoming professionalised, respect for clearly demarcated tasks in their contracts. Citing Rancière (2001), Lepetitcorps (2018: 92–93) comments that these women belong to three groups (women, domestic workers, foreigners) which the state had traditionally excluded from citizenship, and that their activism in fighting for their rights has generated a new political subjectivity in which they brought a private issue into the public domain.

At the same time, one needs to recognise the intersectionality of these activisms for as Kudakwashe (2019: 30), in relation to Zimbabwean domestic worker activism in South Africa, argues 'any intervention or mobilisation that does not take intersectionality into account cannot redress the specific manner in which they are subordinated. . . . African women do not constitute a homogenous category politically or otherwise and do not necessarily share or perceive "objective" gender interests as they are both united and divided by ethnicity and nationality'.

There are few systematic studies of migrant associations and networks that focus on women's and gendered issues at national levels. In Ireland, De Tona and Lentin (2011) identified 40 women's groups in Ireland in the first decade of the century finding that they tended to have very loose multi-national, multi-ethnic and multi-faith boundaries and that most of the networks were inclusive and expanding, usually avoiding the more traditional hierarchical community structures where networking assumes a gendered act of resistance to communitarian discourses and politics. Some of the key associations have played major roles in challenging and transforming politics. For example, AkiDwA (African and Migrant Women's Network) worked to mitigate the shift from a jus soli to a jus sanguinis citizenship policy (removing automatic Irish citizenship from children born in Ireland) in 2005 and campaigning to have migrant mothers have the right of residence to care for their children which had been removed in 2003. It was restored in 2005, 6 months after the change in citizenship laws. This network has also been involved in issues of gender-based violence and black women in the labour market. Another organisation of Muslim women (NOUR), set up by an Algerian woman who entered through family reunification, sought to enable women to debate the nature of Islam in Ireland, advocate for gendered services for women and destabilise stereotypes of Muslim women. Since the 2008 financial crisis and the growing privatisation of welfare, competition, commodification and the struggle for resources have made it more difficult for NGOs and networks to survive and engage in solidarity, as a comparative study of France and Britain highlights (Bassel & Emejulu, 2018).

Finally, wider international networks of migrant activism can shape more transnational and ideally global efforts of mobilisation for migrant justice. For instance, the *European Network of Migrant Women* (https://www.migrantwomennetwork. org/) is a feminist secular migrant-women led platform of NGOs and individual women that advocates for the rights, freedoms and dignity of migrant, refugee and ethnic minority women and girls in Europe. Their membership ranges from grassroots service providers to NGOs focused on advocacy and research. Members cover a diverse range of subjects in the area of human rights of migrant and refugee women, with economic empowerment, anti-discrimination and access to justice and combatting 'Male Violence against Women and Girls', being the most frequent activities.

Various analyses (Lahusen & Theiss, 2019) show that most solidarity organisations remain active primarily at the local and/or national level/s, and that only a minority of solidarity organisations are engaged in cross-national activities. Transnational activism entails a web of transnational partners, organisations and activities for a politicised mission which will lead to more global activism. It is important that such cross-national organised activisms start developing at the grass-roots level where there is no direct dependence on supra-national and inter-governmental governance of organisational linkages while adhering to more organic forms of organising activisms.

6.4   Conclusion

As we have seen in this chapter, the notion of integration as a form of policy intervention has been subjected to increasing academic critique. Whilst its links with the transposition of colonial governance to settler societies has been highlighted, its historical embeddedness in gendered dimensions of othering of non-Western populations, also warrants our attention. It has been extensively argued that the trope of women portrayed as vulnerable, in a state of victimhood and in need of protection (see Chap. 5) is one aligned to their integration seen as both problem and solution for the project of migrant integration at large. Frequently, the discussion of socio-economic aspects of integration policies are overlain with discourses focusing on challenges with integrating Muslim women, thereby shifting the debates to ethnoreligious undertones with significant policy exclusions. In the past 20 years or so, Muslim women have been propelled to the foreground of integration debates and their bodies a battleground over which national identity and security are fought. Gendered perspectives on integration discourses thus require a more nuanced intersectional lens to understand how integration measures affect the diversity of family and labour migrants and refugees. Mechanisms of the hostile environment function as assemblages that ignore such intersectional situatedness in policy and thus exacerbate exclusions.

In seeking to go beyond integration discourses, migrant associations and networks have sought to resist the exclusion of migrant and refugee women from citizenship, extend their socio-economic rights and challenge stereotypical images. Political subjectivities of rights claims are thus political possibilities for inclusion underpinned by those progressive struggles that understand the politics of resistance as the politics for belonging. This is occurring at national, European and international levels though the degree to which rights-based instruments are able to effect change and improved conditions depend on local and national contexts, as can be seen in the application of the decent work for domestic workers agenda.

References

Alba, R., & Foner, N. (2015). *Strangers no more: The challenges of integration in North America and Western Europe*. Princeton University Press.

Amelina, A., & Faist, T. (2012). De-naturalizing the national in research methodologies: Key concepts of transnational studies in migration. *Ethnic and Racial Studies, 35*(10), 1707–1724.

Anthias, F. (2013). Moving beyond the Janus face of integration and diversity discourses: Towards an intersectional framing. *The Sociological Review, 61*(2), 323–343.

Anthias, F., & Pajnik, M. (Eds.). (2014). *Contesting integration, engendering migration*. Palgrave.

Anthias, F., Kontos, M., & Morokvacic-Müller, M. (Eds.). (2013). *Paradoxes of integration: Female migrants in Europe*. Springer.

Bassel, L. (2021). Gender, nationalism and deserving citizenship. In C. Mora & N. Piper (Eds.), *The Palgrave handbook of gender and migration* (pp. 475–490). Palgrave Macmillan.

Bassel, L., & Emejulu, A. (2018). *Minority women and austerity: Survival and resistance in France and Britain*. Policy Press.

Bech, E., Borevi, K., & Mouritsen, P. (2017). A 'civic turn' in Scandinavian family migration policies? Comparing Denmark, Norway and Sweden. *Comparative Migration Studies, 5*(7). https://doi.org/10.1186/s40878-016-0046-7

Birkvad, S. R. (2019). Immigrant meanings of citizenship: Mobility, stability, and recognition. *Citizenship Studies, 23*(8), 798–814.

Bissenbakker, M. (2019). Attachment required: The affective governmentality of marriage migration in the Danish Aliens Act 2000–2018. *International Political, 13*(1). https://doi.org/10.1093/ips/olz001

Bonizzoni, P. (2018). Policing the intimate borders of the nation: A review of recent trends in family-related forms of immigration control. In J. Mulholland, N. Montagna, & E. Sanders-McDonagh (Eds.), *Gendering nationalism. Intersections of nation, gender and sexuality* (pp. 223–239). Palgrave Macmillan.

Bonjour, S., & de Hart, B. (2013). A proper wife, a proper marriage. Constructions of 'us' and 'them' in Dutch family migration policy. *European Journal of Women's Studies, 20*(1), 61–76.

Bonjour, S., & Duyvendak, J. W. (2018). The "migrant with poor prospects": Racialized intersections of class and culture in Dutch civic integration debates. *Ethnic and Racial Studies, 41*(5), 882–900.

Bonjour, S., & Kraler, A. (2015). Introduction: Family migration as an integration issue? Policy perspectives and academic insights. *Journal of Family Issues, 36*(11), 1407–1432.

Briddick, C. (2020). Precarious workers and probationary wives: How immigration law discriminates against women. *Socio and Legal Studies, 29*(2), 201–224.

Brubaker, R. (2001). The return of assimilation? Changing perspectives on integration and its sequels in France, Germany and the United States. *Ethnic and Racial Studies, 24*(4), 531–548.

Castañeda, E. (2018). *A place to call home: Immigrant exclusion and urban belonging in New York, Paris, and Barcelona.* Stanford University Press.

Chang, H. (2021, January 1). *The headscarf debate in Germany. Taking it on or taking it off?* European Academy on Religion and Society.

Chau, H. S., Pelzelmayer, K., & Schwiter, K. (2018). Short-term circular migration and gendered negotiation of the right to the city: The case of migrant live-in care workers in Basel, Switzerland. *Cities, 76*, 4–11.

Cherubini, D., Garofalo Geymonat, G., & Marchetti, S. (2018). Global rights and local struggles. The case of ILO Convention N.189 on domestic work. *The Open Journal of Sociopolitical Studies, 11*(3), 717–742.

Cox, S. (2018, September 21). *Case watch: A victory in Europe for Muslim women's right to wear a headscarf.* Open Society. https://www.justiceinitiative.org/voices/case-watch-victory-europe-muslim-women-s-right-wear-headscarf

Dahinden, J. (2016). A plea for the 'de-migranticization' of research on migration and integration. *Ethnic and Racial Studies, 39*(13), 2207–2225.

Dahinden, J., & Manser-Egli, S. (2021, March 3). *Gendernativism in the (Il)liberal state: The burqa ban in Switzerland.* https://nccr-onthemove.ch/blog/gendernativism-in-the-illiberal-state-the-burqa-ban-in-switzerland/

de Haas, H., Castles, S., & Miller, M. (2019). *The age of migration: International population movements in the modern world.* Palgrave Macmillan.

De Tona, C., & Lentin, R. (2011). 'Building a platform for our voices to be heard': Migrant women's networks as locations of transformation in the Republic of Ireland. *Journal of Ethnic and Migration Studies, 37*(3), 485–502.

Eggbø, H. (2010). The problem of dependency: Immigration, gender and the welfare state. *Social Politics, 17*(3), 295–322.

Eggbø, H., & Brekke, J. P. (2019). Family migration and integration. The need for a new research agenda. *Nordic Journal of Migration Research, 9*(4), 425–444.

European Commission. (2019). *EU court rules Danish 'attachment' requirement for family reunification unlawful in Turkish cases.* https://ec.europa.eu/migrant-integration/news/eu-court-rules-danish-attachment-requirement-for-family-reunification-unlawful-in-turkish-cases. Accessed 28 Aug 2020.

European Migration Network. (2017). *Family reunification of third country nationals in the EU plus Norway: National practices.*

Favell, A. (2003). Integration nations: The nation-state and research on immigrants in Western Europe. *Comparative Social Research Yearbook, 22*(November), 13–42.

Favell, A. (2019). Integration: Twelve propositions after Schinkel. *Comparative Migration Studies, 7*(21). https://doi.org/10.1186/s40878-019-0125-7

Goodman, S. (2011). Controlling immigration through language and country knowledge requirements. *West European Politics, 34*(2), 235–255.

Hagelund, A. (2020). After the refugee crisis: Public discourse and policy change in Denmark, Norway and Sweden. *Comparative Migration Studies, 8*(13). https://doi.org/10.1186/s40878-019-0169-8

Hancock, C. (2015). The republic is lived with an uncovered face: (Un)dressing French citizens. *Gender, Place and Culture, 22*(7), 1023–1040.

Hancock, C., & Mobillon, V. (2019). "I want to tell them, I'm just wearing a veil, not carrying a gun!" Muslim women negotiating borders in femonationalist Paris. *Political Geography, 69*, 1–9.

Hancock, C., Blanchard, S., & Chapuis, A. (2018). Banlieusard.e.s claiming a right to the city of light: Gendered violence and spatial politics in Paris. *Cities, 76*, 23–28.

Harvey, D. (2008). The right to the city. *New Left Review, 53*, 23–40.

Hernes, V. (2018). Cross-national convergence in times of crisis? Integration policies before, during and after the refugee crisis. *West European Politics, 41*(6), 1305–1329. https://doi.org/10.1080/01402382.2018.1429748

Home Office. (2011a). *Family migration: A consultation.* Home Office.

Home Office. (2011b). *Family migration: Evidence and analysis.* Home Office.

Isaakyan, I., & Triandafyllidou, A. (2014). Anglophone marriage-migrants in southern Europe: A study of expat nationalism and integration dynamics. *International Review of Sociology, 24*(3), 374–390.

Joppke, C. (2007). Beyond national models: Civic integration policies for immigrants in Western Europe. *West European Politics, 30*(1), 1–22.

Joppke, C. (2009). *Veil.* Mirror of identity.

Kallio, K. P., Häkli, J. & Bäcklund, P. (2015). Lived citizenship as the locus of political agency in participatory policy. *Citizenship Studies, 19*(1), 101–119.

Kallio, K. P., Wood, B. E., & Häkli, J. (2020). Lived citizenship: Conceptualising an emerging field. *Citizenship Studies, 24*(6), 713–729.

Karayianni, C., & Christou, A. (2020). Feminisms, gender and social media: Public and political performativities regarding sexual harassment in Cyprus. *Feminist Encounters: A Journal of Critical Studies in Culture and Politics, 4*(2). https://doi.org/10.20897/femenc/xxxx

Kirk, K. (2010). *A gendered story of citizenship: A narrative analysis of Dutch civic integration policies.* PhD thesis Queens University Belfast.

Klarenbeek, L. (2019). Reconceptualising 'integration' as a two-way process. *Migration Studies,* mnz033. https://doi.org/10.1093/migration/mnz033

Kofman, E., Saharso, S., & Vacchelli, E. (2015). Gendered perspectives on integration measures. *International Migration, 53*(4), 77–89.

Kontos, M. (2014). Restrictive integration policies and the construction of the migrant as 'unwilling to integrate': The case of Germany. In F. Anthias & M. Pajnik (Eds.), *Contesting integration, engendering migration* (pp. 125–142). Palgrave Macmillan.

Korteweg, A. (2017). The failures of 'immigrant integration': The gendered racialized production of non-belonging. *Migration Studies, 5*(3), 428–444.

Korteweg, A., & Triadafilopoulos, T. (2013). Gender, religion, and ethnicity: Intersections and boundaries in immigrant integration policy making. *Social Politics, 20*(1), 110–129.

Korteweg, A., & Yurdakul, G. (2020). Liberal feminism and postcolonial difference: Debating headscarves in France, the Netherlands, and Germany. *Social Compass.* https://doi-org.ezproxy.mdx.ac.uk/10.1177/0037768620974268

Kostakopoulou, D. (2014). The anatomy of civic integration. In F. Anthias & M. Pajnik (Eds.), *Contesting integration, engendering migration* (pp. 37–63). Palgrave Macmillan.

Kudakwashe, V. (2019). *Zimbabwean migrant domestic worker activism in South Africa.* Working Paper 55. https://opendocs.ids.ac.uk/opendocs/handle/20.500.12413/14871. Last accessed 5 Sept 2020.

Lahusen, C., & Theiss, M. (2019). European transnational solidarity: Citizenship in action? *American Behavioral Scientist., 63*(4), 444–458.

Lechner, C. (2011). *Perception and impact of pre-entry tests for TCN's.* EFMS, Bamberg: PROSINT German Country Report.

Lefebvre, H. (1996). *Writings on cities* (E. Kofman & E. Lebas, Trans.). Blackwell.

Lepetitcorps, C. (2018). Migrant women in trade unions. Domestic service activism in France. In M. Amrith & N. Sarahoui (Eds.), *Gender, work and migration: Agency in gendered labour settings* (pp. 83–98). Routledge.

Levitt, P. (2012). What's wrong with migration scholarship? A critique and a way forward. *Identities: Global Studies in Culture and Power, 19*(4), 493–500.

Meissner, F., & Heil, T. (2020). Deromanticising integration: On the importance of convivial disintegration. *Migration Studies*, 1–19. https://doi.org/10.1093/migration/mnz056

Mepschen, P., Duyvendak, W., & Tonkens, E. (2010). Sexual politics, orientalism and multicultural citizenship in the Netherlands. *Sociology, 44*, 962–979.

Ministry of Integration. (2002). L152: Proposal for law amending the Aliens Act and the Marriage Act.

Mulally, S. (2015). Introduction. Decent work, domestic work: Gendered borders and limits. In S. Mulally (Ed.), *Care, migration and human rights* (pp. 1–10). Routledge.

Murphy, C. (2015). Access to justice for undocumented migrant workers in Europe: The consequences of constructed illegality. In S. Mulally (Ed.), *Care, migration and human rights* (pp. 110–130). Routledge.

OECD (Organisation for Economic Co-operation and Development). (2017). *Making integration work. family migrants.* OECD. https://www.oecd.org/migration/making-integration-work-9789264279520-en.htm

Olwig, F. (2011). Integration: Migrants and refugees between Scandinavian welfare societies and family relations. *Journal of Ethnic and Migration Studies, 37*(2), 179–196.

Open Society. (2018). *Restrictions on Muslim women's dress in the 28 EU Member States. Current law, recent legal developments and the state of play.* Open Society.

Parrilli, M. D., Montresor, S., & Trippl, M. (2019). A new approach to migrations: Communities-on-the-move as assets. *Regional Studies, 53*(1), 1–5.

Penninx, R. (2019). Problems of and solutions for the study of immigrant integration. *Comparative Migration Studies, 7*(13). https://doi.org/10.1186/s40878-019-0122-x

Phillips, A., & Saharso, S. (2008). The rights of women and the crisis of multiculturalism. *Ethnicities, 8*(3), 291–301.

Prins, W., & Saharso, S. (2008). In the spotlight: A blessing and curse for immigrant women in the Netherlands. *Ethnicities, 8*(3), 365–384.

Raimondi, V. (2019). For 'common struggles of migrants and locals'. *Migrant Activism and Squatting in Athens, Citizenship Studies, 23*(6), 559–576.

Rancière, J. (2001). Citoyennété, culture et politique. In M. Elbaz & D. Helly (Eds.), *Mondialisation, citoyennété et multiculturalisme* (pp. 55–68). L'Harmattan.

Razack, S. H. (2004). Imperilled Muslim women, dangerous Muslim men and civilised Europeans: Legal and social responses to forced marriages. *Feminist Legal Studies, 12*, 129–174.

Renvik, T. A., Manner, J., Vetik, R., Sam, D. L., & Jasinskaja-Lahti, I. (2020). Citizenship and socio-political integration: A person-oriented analysis among Russian-speaking minorities in Estonia, Finland and Norway. *Journal of Social and Political Psychology, 8*, 53–77.

Rytter, M. (2019). Writing against integration: Danish imaginaries of culture, race and belonging. *Ethnos.* https://doi.org/10.1080/00141844.2018.1458745

Samad, Y. (2010). Forced marriages among men: An unrecognized problem. *Critical Social Policy, 30*(2), 189–207.

Schinkel, W. (2017). *Imagined societies: A critique of immigrant integration in Western Europe.* Cambridge University Press.

Schinkel, W. (2018). Against 'immigrant integration': For an end to neocolonial knowledge production. *Comparative Migration Studies, 6*, 31. https://doi.org/10.1186/s40878-018-0095-1

Schinkel, W. (2019). Migration studies: An imposition. *Comparative Migration Studies, 7*, 32. https://doi.org/10.1186/s40878-019-0136-4

Schinkel, W., & van Houdt, F. (2010). The double helix of cultural assimilation and neo-liberalism: Citizenship in contemporary governmentality. *The British Journal of Sociology, 61*(4), 696–715.

Schmidt, G. (2011). Law and identity: Transnational arranged marriages and the boundaries of Danishness. *Journal of Ethnic and Migration, 37*(2), 257–275.

Scholten, P., et al. (2012). *Integration from abroad? Perception and impacts of pre-entry tests for third country nationals. PROSINT comparative report.* ICMPD.

Schultz, C. (2020). A prospect of staying? Differentiated access to integration for asylum seekers in Germany. *Ethnic and Racial Studies, 43*(7), 1246–1264.

Schwenken, H. (2017). The emergence of an impossible movement domestic workers organize globally. In D. Gosewinkel & D. Rucht (Eds.), *Transnational social movements* (pp. 205–228). Berghan Books.

Scott, J. (2007). *The politics of the veil.* Princeton University Press.

Sirriyeh, A. (2015). All you need is love and £18,600': Class and the new UK family migration rules. *Critical Social Policy, 35*(2), 228–247.

Staver, A. (2015). Hard work for love. The economic drift in Norwegian family immigration and integration policies. *Journal of Family Issues, 36*(11), 1453–1471.

Sumption, M., & Vargas-Silva, C. (2019). Love is not all you need: Income requirement for visa sponsorship of foreign family members. *Journal of Economics, Race, and Policy, 2*, 62–76.

Trimikliniotis, N. (2014). The only thing I like integrated is my coffee: Dissensus and migrants integration in the era of Euros-Crisis. In F. Anthias & M. Pajnik (Eds.), *Contesting integration, engendering migration* (pp. 64–85). Palgrave.

Vacchelli, E., & Kofman, E. (2018). Towards an inclusive and gender right to the city. *Cities, 76*, 1–3.

Van Kerckem, K., van Bracht, B., van Putte, P., & Stevens, A. (2013). Transnational marriages on the decline: Explaining changing trends in partner choice among Turkish Belgians. *International Migration Review, 47*(4), 1006–1038.

Van Walsum, S. (2012). *Intimate strangers.* Inaugural lecture Professor of Migration, Law and Family Ties, The Faculty of Law, VU University Amsterdam.

Wagner, R. (2015). Family life across borders: Strategies and obstacles. *Journal of Family Issues, 36*(11), 1509–1528.

Weaver, M. (2018). Burqa bans, headscarves and veils: Timeline of legislation in the west. *The Guardian*, 31 May https://www.theguardian.com/world/2017/mar/14/headscarves-and-muslim-veil-ban-debate-timeline

Wood, B. E. (2013). Young people's emotional geographies of citizenship participation: Spatial and relational insights. *Emotion, Space and Society, 9*, 50–58.

Wray, H., Kofman, E., & Simic, A. (2021). 'Subversive citizens: Using free movement law to bypass the UK's rules on marriage migration. *Journal of Ethnic and Migration Studies, 47*(2), 447–463.

Yurdakul, G., & Korteweg, A. (2013). Gender equality and immigrant integration: Honor killing and forced marriage debates in the Netherlands, Germany, and Britain. *Women's Studies International Forum, 41*, 204–214.

Yuval-Davis, N., Wemyss, G., & Cassidy, K. (2018). Everyday bordering, belonging and the reorientation of British immigration legislation. *Sociology, 52*(2), 228–244.

Chapter 7
Conclusion

At the end of a short journey, we can attest to the flourishing production of knowledge on gender and migration that has built up over the past 30 years in particular. Though we have on the whole referred to works in English, there is an extensive literature in other major languages, such as French, German, Italian and Spanish which have emerged from different social science traditions, in recognition of the significance of gendered migrations and feminist movements. English has come to dominate writing in this field (Kofman, 2020), ironically in large part through the European funding of comparative research as well as transatlantic exchanges (Levy et al., 2020). The past 20 years have been a rapid period of intellectual exchange in this field through networks and disciplinary associations, such as the International and European Sociological Associations or IMISCOE which supported a cluster on Gender, Generation and Age (2004–2009). The IMISCOE Migration Research Hub (https://www.migrationresearch.com/) demonstrates the extensive production on gender issues and their connections with other theories and fields of migration. The economic and social transformations brought about by globalisation and transnationalism, and how its unequal outcomes and identities need to be understood through an intersectional lens (Amelina & Lutz, 2019), have heavily shaped studies of gender and migration (see Chap. 2). Indeed intersectionality has been suggested by some as the major contribution of contemporary feminism to the social sciences, and, has certainly been a theoretical insight that has travelled widely and rapidly from the Anglo world to Europe (Davis, 2020; Lutz, 2014) since it was defined by Kimberlé Crenshaw (1989). We should, however, also remember that it had antecedents in the writing of anti-racist feminists on racist ideology and sex by the French sociologist Claude Guillaumin (1995), on the trinity of gender, race and class in the UK (Anthias & Yuval-Davis, 1992; Parmar, 1982) and by scholars in Australia (Bottomley et al., 1991) and Canada (Stasiulis & Yuval-Davis, 1995).

What the application of an intersectional approach has brought to the fore are the complex and intersecting inequalities in the experiences and outcomes of migration, reinforced by restrictive borderings and categorisations generated by immigration

© The Author(s) 2022   117
A. Christou, E. Kofman, *Gender and Migration*, IMISCOE Research Series,
https://doi.org/10.1007/978-3-030-91971-9_7

policies regulating labour, family and asylum flows. The three main categories of intersectional analysis have been gender, race and class. The latter is often poorly captured, though there has been growing attention to how migration shapes class positions transnationally, for example in the concept of contradictory class mobility (Parreñas, 2001) and class differences among migrants of the same nationality (Horst et al., 2016). New social divisions have also been incorporated into intersectional approaches. Amongst the most significant are men and masculinities, sexualities, age (youth and older migrants) and, to a lesser extent, disability (Fiske & Giotis, 2021). As we saw in Chap. 2 in our discussion of the shift from women to gender, the latter often continues to be reduced to women. The tendency to focus on women and a call for a more gendered approach has been prominent in the critical discussion of the application of the concept of vulnerability in migration governance and humanitarian management (see Chap. 5). These divisions intersect with immigration and integration measures and policies (see Chap. 6). Though supposedly gender neutral, immigration policies have profound gender implications through their conceptualisation of the deserving and the undeserving in relation to entry, right to residence and citizenship (Boucher, 2016; Kofman & Raghuram, 2015; Stasiulis, 2020).

As Stasiulis et al. (2020) underscore, the analytical gains when deploying an intersectional lens have to do with making apparent the oppression, violence, discrimination and dehumanisation of specific migrant groups. This focus draws attention to the interconnected dynamics of power structures and agonisingly reveals that the relationship between migration and social injustice continues to be historically and contemporaneously a phenomenon of social erasure within states, policies, laws and social consciousness. Linking back to Chap. 6, these are intertwining resource, not just for scholars who seek to understand them, but also for activists who wish to transform and eradicate inequities.

We have highlighted the development of intersectionality among migration scholars and would argue that adopting an historical perspective, including the role of colonialism and how categories of gender, sexuality and race were constructed during colonial modernity (Mayblin & Turner, 2021; Chap. 7), is important in acquiring a better understanding of how particular topics and approaches have emerged and evolved. Throughout the book we have sought to draw out the changing theorisations and approaches to gender and migration as a whole (Chaps. 1 and 2) and in relation to specific forms of migration (Chaps. 3, 4, and 5) and participation (Chap. 6). The feminisation of migration provides a good example of the need to place trends in the *longue durée* and questions the idea that the process has consisted of a linear progression. Equally important is the geographical dimension highlighting considerable variations between localities, regions and states. One often hears that domestic and care work is being performed largely by migrants, but a more detailed picture becomes apparent in the analysis of major metropolitan and other areas. It is in metropolitan centres that migrant labour in these sectors is the highest (Kofman & Raghuram, 2015) whereas in rural areas and regions without a strong history of immigration, the work is still undertaken by working class women (Howard and Kofman, 2020). As Glick Schiller and Çaglar

(2009) point out, we must be careful about extracting from particular localities, especially in major cities where most migration research takes place, to the national level.

Arriving at the destination of our collaborative writing journey and completion of this book has culminated with three key critical moments: 'COVID-19', an unprecedented new pandemic with its devastating impact on the loss of lives, shattering of economies, gendered inequalities in social reproduction and transformation of lifestyles and mobilities; the surrealistic, in our view, outcome of the 'Brexit' referendum in the UK and its aftermath; and, the intellectual and political consequences of the 'Black Lives Matter' protests in the United States and globally.

In terms of the **COVID-19 pandemic**, Alan Gamlen (2020) contemplated the fate of migration and mobility after it ends and posed ten key questions about future transformations. These range from the future of labour migration, migrant decision-making, anti-immigrant sentiments and autocratic regimes, migration restrictions, international student migration etc. Above all Gamlen asked whether we are witnessing the end of the age of migration (Castles et al., 2014), at least temporarily. We have seen unprecedented closures of borders, even within countries and a radical reduction of mobility, including tourism, and migration. Some have suggested, in line with current trends in major immigration countries, of privileging the skilled and restricting the lesser skilled to strictly temporary periods of residence which could approximate to a Singapore model and, what Stasiulis (2020) notes for Canada, as the disposability of the less skilled. As we saw in Chap. 3, the distinction between the skilled and the less skilled, has clear implications for labour markets and migrant rights.

Migrant domestic care workers, who often already have less labour rights and social protection, have frequently been called upon to provide the essential work to sustain households and societies (Rao et al., 2021). A report (Leiblfinger et al., 2020) on the impact of the pandemic on live-in care workers in Germany, Austria and Switzerland highlights the fact that the pandemic had extensively reduced circular migration of live-in care workers between their home countries in Central and Eastern Europe and their live-in care worker residences. The report offered insights into the impact of travel restrictions during the pandemic and transnational live-in care. While differences are identified between countries, the authors draw on similarities of the impact of such measures in reducing the interests of migrant care workers in comparison to care receivers. We can infer that the pandemic will not probably lead to positive changes to the working conditions of migrant carers or immigration policies after the initial championing of essential and key workers. At the same time, the pandemic has also exposed Western Europe's reliance on seasonal Eastern European migrants for other parts of the economy, with future border closures certainly impacting on economic stability (Kondan, 2020). At the time of submitting the final manuscript in October 2021, European states have begun to emerge from the second lockdown and benefitted from widespread vaccination but 80% of vaccines have gone to upper and upper middle income countries, with poorer countries in Africa in particular lacking access to vaccines and having less ability to sustain lockdowns (https://ourworldindata.org/covid-vaccinations). It is also not clear what will emerge in relation to gender inequalities relating to gender violence, unemployment, changes in the labour market and the additional burden of social reproduction in the home.

The future of diaspora engagement will also be an important avenue shaping the future of nation-states and their transnational networks as they have again been disrupted by the pandemic and travel restrictions (ibid). Nevertheless, diasporas have proven that they can serve as 'legitimate' actors during current periods of crises leading to more efficient policy implementation at both local and transnational level (Dag, 2020).

Travel restrictions and post-pandemic changes in organising social and research life will most likely see an increase in the shift to 'digital' research in migration studies. This will have methodological, analytical and ethical implications but also opportunities in perhaps reaching more groups and gendering more of our migration research, while being more in tune with intersectional implications of this work. This direction might unveil more 'digital passages' (Koen, 2015) in capturing migration processes which involve digital identity construction, transnational caring arrangements which involve online provision of migrant care (Janta & Christou, 2019) and the negotiation of gendered diaspora and generational cultural expectations.

Our second major event impacting on European mobility migration patterns is that of **Brexit** which was voted in favour of by a narrow majority of 51.9% in a referendum held on 23 June 2016. After the end of the transition period in January 2021, a new immigration policy has come into force to reflect the UK's withdrawal from the EU and its free movement policies. The vote reflected an imperial nostalgia and global reach, in which Britannia ruled the waves (Agnew, 2020) and an Englishness 'reasserted through a racializing, insular nationalism' (Virdee & McGeever, 2018: 1804). The referendum has already had the effect of reducing EU immigration and encouraging those already settled to leave. A sharp distinction between the skilled and the less skilled based on income levels has been imposed as well as an accrued control of migrants in which future EU and non-EU migrants are both subjected to the hostile environment of everyday bordering practices (housing, health, education, deportation). As we saw in Chap. 3, high income favours men in terms of eligibility for immigration. According to the UK Institute for Public Policy Research (Morris, 2020) under current immigration proposals, 36% of men would be eligible for a skilled visa but only 26% of women. 59% of construction workers and 66% of the people currently working in the health and social care sector would not be eligible.

Thus care labour is likely to be particularly hard hit. For immigration policy the value of one's labour is equated with the price of it, and given that care remains discounted and under-valued, it not only fails the entry level income criteria but also has not been given any special consideration as a shortage occupation. It is not clear whether those in need of care will be left with poorer quality care or, as commonly happens, families, and in this instance, largely women, will be left to care for members of their family. As has been evident during the COVID-19 pandemic, women, especially those with children (Fisher & Ryan, 2021), have assumed, to an even greater extent than previously, caring responsibilities. At the same time, restrictions imposed by COVID-19 have, for the time, being delayed by the gendered impact of new immigration policies for the entry of workers. However, new post Brexit regulations will also have implications for the right to family life as future EU migrants will have to comply with extremely high minimum income levels (see Chap. 4) to bring in spouses and children as they cease to benefit from the more generous European freedom of movement. They will once again return to the status of mobile workers rather than fellow citizens (D'Angelo & Kofman, 2018). It is likely to have a negative impact on student flows and on youth mobility as the UK has imposed expensive international student fees on EU students and withdrawn from the EU Erasmus+ scheme.

The third event of **Black Lives Matter** is yet another reminder that any migrant crisis is a racial crisis (De Genova, 2018; Kirtsoglou & Tsimouris, 2018). As Bridget Anderson (2020) argues, migration studies in the past 30 years has drifted apart from race and ethnic studies in the UK, although there are scholars who have tried to bring them together. While immigration policies are no longer as blatantly racist as they were in the past, especially in settler societies, it operates in part through the more restrictive economic criteria favouring the skilled through a dynamic geopolitical landscape of centre and periphery. China and India, are now two of the major sources of skilled migrants. Even so, these nationalities, may also face the visible and invisible walls of white privilege in accessing professional employment and discrimination in the workplace in the country of destination (Carangio et al., 2021). Racism is most forceful in the application of immigration regulations, for example in detention and deportation where Black Lives can be discarded and are equated with being a migrant whose belonging is questioned (Anderson, 2020). Many have suggested we need a better historical education about immigration to bring out the effects of slavery and colonialism (Mayblin & Turner, 2021; Yeo, 2020). We also need to recognise the role of Islamophobia in immigration and integration policies and its gendered representations of migrants and refugees (see Chap. 6). The very mobility of these populations as well as Roma, Gypsies and Travellers, who have experienced some of the most systematic racism in Europe, also have to be included in our migration scholarship of the past, present and future in our curricula and research.

These recent and momentous developments will undoubtedly have an important impact on both the nature of migration and mobility globally, and, will need to be engaged with critically from gendered and intersectional perspectives. They will be part of the continuing and lively debates that we have shown characterise writings on gender and migration and efforts to take forward social justice initiatives based on insightful critiques.

References

Agnew, J. (2020). Taking back control? The myth of territorial sovereignty and the Brexit fiasco. *Territory, Politics, Governance, 8*(2), 259–272.

Amelina, A., & Lutz, H. (2019). *Gender and migration. Transnational and intersectional perspectives*. Routledge.

Anderson, B. (2020). *Black lives matter – Whatever their nationalities*. Migration and Mobilities Bristol. https://migration.blogs.bristol.ac.uk/2020/06/30/black-lives-matter-whatever-their-nationality/

Anthias, F., & Yuval-Davis, N. (1992). *Racialized boundaries: Race, nation, gender, colour and class and the anti-racist struggle*. Routledge.

Bottomley, G., de Lepervanche, M., & Martin, J. (Eds.). (1991). *Intersexions: Gender/class/culture/ethnicity*. Allen & Unwin.

Boucher, A. (2016). *Gender. Migration and the global race for talent*. Manchester University Press.

Carangio, V. K., Farquharson, S. B., & Rajendran, D. (2021). Racism and White privilege: Highly skilled immigrant women workers in Australia. *Ethnic and Racial Studies, 44*(1), 77–96.

Castles, S., Miller, M., & de Haas, H. (2014). *Age of migration*. Palgrave Macmillan.

Crenshaw, K. (1989). Demarginalizing the intersection of race and sex: A black feminist critique of antidiscrimination doctrine, feminist theory and antiracist politics. *University of Chicago Legal Forum, 1*, 139–167.

D'Angelo, A., & Kofman, E. (2018). 'From mobile worker to fellow citizen and back again? The future status of EU citizens in the UK' social. *Policy and Society, 24*(2), 331–343.

Dag, V. (2020, May 8). *In times of crisis, diaspora groups know what to do*. Open Democracy. https://www.opendemocracy.net/en/north-africa-west-asia/times-crisis-diaspora-groups-know-what-do/

Davis, K. (2020). Who owns intersectionality? Some reflections on feminist debates on how theories travel. *European Journal of Women's Studies, 28*(2), 1113–1127.

De Genova, N. (2018). The "migrant crisis" as racial crisis: Do *Black Lives Matter* in Europe? *Ethnic and Racial Studies, 41*(10), 1765–1782.

Fisher, A., & Ryan, M. (2021). Gender inequalities during Covid-19. *Group Processes and Intergroup Relations, 24*(2), 237–245.

Fiske, L., & Giotis, C. (2021). Refugees, gender and disability: Examining interactions through refugee journeys. In C. Mora & N. Piper (Eds.), *The Palgrave handbook of gender and migration* (pp. 441–454). Palgrave Macmillan.

Gamlen, A. (2020). *Migration and mobility after the 2020 pandemic: The end of an age?* WP 20-146, Centre on Migration, Policy and Society, University of Oxford.

Glick Schiller, N., & Çaglar, A. (2009). Towards a comparative theory of locality in migration studies: Migrant incorporation and the city scale. *Journal of Ethnic and Migration Studies, 35*(2), 177–202.

Guillaumin, C. (1995). *Racism. Power and ideology. Sexism*. Routledge.

Horst, C., Pereira, S., & Sheringham, O. (2016). The impact of class on feedback mechanisms: Brazilian migration to Norway, Portugal and the United Kingdom. In O. Bakewell, G. Engbersen, M. L. Fonseca, & C. Horst (Eds.), *Beyond networks. Migration, diasporas and citizenship* (pp. 90–112). Palgrave Macmillan.

Janta, H., & Christou, A. (2019). Hosting as social practice: Gendered insights into contemporary tourism mobilities. *Annals of Tourism Research, 74*, 167–176.

Kirtsoglou, E., & Tsimouris, G. (2018). Migration, crisis, liberalism: The cultural and racial politics of islamophobia and "radical alterity" in modern Greece. *Ethnic and Racial Studies, 41*(10), 1874–1892.

Koen, L. (2015). *Digital passages: Migrant youth 2.0: Diaspora, gender and youth cultural intersections*. Amsterdam University Press.

Kofman, E. (2020). Unequal internationalisation and the emergence of a new epistemic community: Gender and migration. *Comparative Migration Studies, 8*, 36. https://doi.org/10.1186/s40878-020-00194-1

Kofman, E., & Raghuram, P. (2015). *Gendered migrations and global social reproduction*. Palgrave Macmillan.

Kondan, S. (2020, August 25). *Southeastern Europe looks to engage its diaspora to offset the impact of depopulation.* Migration Policy Institute. https://www.migrationpolicy.org/article/southeastern-europe-seeks-offset-depopulation-diaspora-ties

Leiblfinger, M., Prieler, V., Schwiter, K., Steiner, J., Benazha, A., & Lutz, H. (2020, May 14). *Impact of the COVID-19 pandemic on live-in care workers in Germany, Austria, and Switzerland.* https://ltccovid.org/2020/05/14/impact-of-the-covid-19-pandemic-on-live-in-care-workers-in-germany-austria-and-switzerland/

Levy, N., Pisarevskaya, A., & Scholten, P. (2020). Between fragmentation and institutionalization: The rise of migration studies as a research field. *Comparative Migration Studies, 8*, 29. https://doi.org/10.1186/s40878-020-00200-6

Lutz, H. (2014). *Intersectionality's (brilliant) career; how to understand the attractiveness of the concept.* Gender, diversity and migration working paper 1. Frankfurt am Main: Goethe University Frankfurt, Department of Social Sciences.

Mayblin, L., & Turner, J. (2021). *Migration studies and colonialism.* Polity Press.

Morris, M. (2020). *Building a post-brexit immigration system for the economic recovery.* Institute for Public Policy.

Parmar, P. (1982). Gender, race and class: Asian women in resistance. In Centre for Contemporary Culture Studies (Ed.), *The empire strikes back: Race and racism in the 70s Britain* (pp. 236–275). Hutchinson & Co.

Parreñas, R. (2001). *Servants of globalization: Women, migration and domestic work.* Stanford University Press.

Rao, S. S., Gammage, J. A., & Anderson, E. (2021). Human mobility, COVID-19, and policy responses: The rights and claims-making of migrant domestic workers. *Feminist Economics.* https://doi.org/10.1080/13545701.2020.1849763

Stasiulis, D. (2020). Elimi(nation): Canada's "post-settler" embrace of disposable migrant labour. *Studies in Social Justice, 14*(1), 22–54.

Stasiulis, D., & Yuval-Davis, N. (1995). *Unsettling settler societies: Articulations of gender, race, ethnicity and class.* Sage.

Stasiulis, D., Jinnah, Z., & Rutherford, B. (2020). Migration, intersectionality and social justice. *Studies in Social Justice, 14*(1), 1–21.

Virdee, S., & McGeever, B. (2018). Racism, crisis, brexit. *Ethnic and Racial Studies, 41*(10), 1802–1819.

Yeo, C. (2020, June 10). *Race, racism and immigration in the United Kingdom: Black lives matter.* Free Movement Blog. https://www.freemovement.org.uk/black-lives-matter/